Physics

A Science in Quest of an Ontology

PHYSICS

A SCIENCE IN QUEST OF AN ONTOLOGY

Wolfgang Smith

Philos-Sophia Initiative Foundation

Series Editor: John Trevor Berger

Hardcover ISBN: 979-8-9851470-3-2
Paperback ISBN: 979-8-9851470-4-9
eBook ISBN: 979-8-9851470-5-6

Library of Congress Control Number: 2022913559

Philos-Sophia Initiative Foundation
www.philos-sophia.org

To my dear friend
Richard Kyle DeLano

Requiescat in pace

TABLE OF CONTENTS

PREFACE

A VERY LONG TIME AGO, when applying to Cornell University as a prospective freshman, the question was put to me on the application form what I wanted to major in, and why. My answer was that I wished to major in physics "because I believe that physics is the key to understanding the universe." Well, I no longer think so, and have in fact become persuaded that the matter stands just the other way round: that in order to "understand physics," one needs first to attain a certain insight concerning the universe: an *ontological* insight, to be precise.

This recognition—or conviction—has moreover come by stages, beginning with my reflections relating to the so-called measurement problem which underlie *The Quantum Enigma*, published in 1995. What finally led to the proffered resolution of that problem—to my satisfaction, at least—is the discernment of an *ontological* distinction between two domains or "levels of being" which are nowadays almost universally identified: the *physical* and what I term the *corporeal*. The crucial fact concerning the act of measurement proves then to be a transition between these two ontological planes—from the physical to the corporeal—which implies (1) that this transition must be *instantaneous*; and (2) that by virtue of this fact it cannot be accomplished by means of the causality upon which physics as such is based. What is called for is a mode of causation that acts instantaneously: *vertical* causation I call it, in contrast to the "horizontal" kind known to physics. To which I would add that, three years after the publication of *The Quantum Enigma*, the existence of VC was confirmed by William Dembski's mathematical discovery to the effect that *horizontal causation cannot produce what is termed "complex specified information,"* a matter to which we shall recur.

It follows that the measurement problem cannot be solved on the grounds of physics *per se*—which may account for the fact that, in what is now close to a century of endeavor involving top physicists, the impasse has not in truth been broken. Claims to the contrary have of course been made: there are authorities, for instance, who maintain that "superdeterminism" is the answer; yet it seems to me that this notion itself is about as problematic as the issue it supposedly resolves.

Such is not the case when it comes to the conceptions of *corporeal* being and *vertical* causation, which are *ontological*, and not *ad hoc*, but grounded in the metaphysical traditions of mankind. It seems to me that metaphysics does perforce enter the cognitive picture, and that our only real choice is restricted to the *kind*: whether it be authentic or spurious. And I incline to believe that the worst metaphysics is generally to be found among those who claim not to have any at all.

We maintain, thus, that when it comes to the foundations of physics, metaphysical considerations enter *de jure*—whether we like it or not, and most importantly, whether we realize it or not. The burgeoning chaos in contemporary physics on its foundational level bears this out: read what expert critics like Sabine Hossenfelder and Alexander Unzicker have to tell us regarding current particle physics and the goings on at CERN—and you will see exactly what I mean.[1]

Getting back to my own involvement with the ontology of physics, it began, as I have said, with the recognition of two ontological principles: the distinction between a *corporeal* object X and the associated *physical* object SX, together with *vertical* causality, an ontological mode of causation needed to transition from SX to X. And so the matter stood until quite recently, when the issue of "wholeness" began to present itself. At that point a third ontological notion came insistently into view: *irreducible wholeness*, I call it. And the more deeply one reflects

1. Cf. Part II, ch. 3. I have also dealt with this issue in *The Vertical Ascent* (Philos-Sophia Initiative Foundation, 2021), ch. 4.

upon the place and function of that IW, the more central and controlling it proves to be.

A stage is reached, in the course of the ensuing reflections, where everything comes together, and one begins to glimpse a previously unsurmised ontological unity and order. One begins in fact to perceive physics in a brand new key; and what I find most striking of all: the distinction between classical physics and quantum theory presents itself now as inherently *ontological*, based upon the presence or absence of an IW. It has thus become apparent why *"no one understands quantum theory,"* as Richard Feynman has famously observed, and what precisely it takes to do so.

It is this fact, above all, that has motivated me to publish these ontological findings separately, which can be done— very perspicaciously—in just three brief chapters. Four closely related articles published on our Philos-Sophia Initiative website in 2021—the period during which the concept of *irreducible wholeness* imposed itself upon me ever more insistently, as *the key to an ontological comprehension of physics*—have been added as Part II.

Wolfgang Smith
June 14, 2022

1

DESCARTES AND THE LOSS
OF THE CORPOREAL

It began, if you will, in the 5th century BC with Democritus of
Abdera when he declared:

> According to vulgar belief, there is color, the sweet and the bit-
> ter; but in reality, only atoms and the void.[1]

Not long thereafter, let us note, his materialist atomism was
attacked by Plato (ca. 427–347) and in effect disqualified. For
the remainder of antiquity—and actually, for about two thou-
sand years—it was regarded as heterodox by the well-informed,
until it reemerged in the seventeenth century to pave the way to
what historians term the Enlightenment.

The central figure in this revival is unquestionably René Descartes, who formalized the metaphysical assumptions of the
Democritean *Weltanschauung* in a single conception of immea-
surable influence: *bifurcation* namely, to use Whitehead's designation. What is being "cut asunder" at one stroke are *res extensae*
or "extended entities" on one side, and *res cogitantes* or "things of
the mind" on the other. This leaves the "real" or objectively existent world enormously reduced and vastly simplified: gone, first
of all, are the innumerable qualities we perceive through our
senses. Gone thus, in a sense, is the world in which we live and

---

1. Hermann Diels, *Die Fragmente der Vorsokratiker* (Dublin: Weid-
mann, 1969), vol II, p. 168.

have our human being; it is replaced, reputedly, by a world of *res extensae* which no human eye has ever seen, and cannot but ultimately reduce to a mechanism of some kind—analogous to those artful clocks in vogue at the time.

The fact is that René Descartes laid the philosophical—or more precisely, the *ontological*—foundations upon which modern physics has been based from its inception to the present day. And that physics made its debut with Sir Isaac Newton's *Principia Mathematica*, published in 1687—less than four decades after the passing of Descartes—and has defined our *Weltanschauung* ever since. By a curious inversion, moreover, which perhaps only a philosopher can discern, that physics and its associated *Weltanschauung* have in turn bestowed a kind of *imprimatur* upon the Cartesian premise itself, enabling it to exert an almost unbreakable stranglehold upon the Occidental intelligentsia at large.

The first major triumph of the Newtonian physics relates to the solar system, which in the Cartesian logic constitutes in itself a gigantic *res extensa*. Newton seemed namely to have substantiated this conjecture by the calculation of planetary orbits, based on two exceedingly simple mathematical laws: the law of motion, which affirms that *force* equals *mass* times *acceleration*, and the law of gravitation, stating that the force of gravitational attraction between two punctual *res extensae* is directly proportional to the product of the masses and inversely proportional to the square of the distance between them. And lo: on the basis of these two formulae—so magnificent in their austere simplicity—it seemed to just about everyone that even the celestial world obeys the newly discovered laws.

In the wake of this initial discovery the victorious sweep of what is nowadays referred to as "classical" physics continued for two centuries, adding numerous and sundry varieties of *res extensae*—for instance, electromagnetic fields—to its inventory. And all went very well up to the year 1897—two hundred ten years following the publication of *Principia*

Mathematica—when a physicist by the name of J. J. Thomson discovered what has subsequently been called the *electron*. What it amounts to, if you will, is the first-ever sighting of a Democritean atom, more than two millennia after the venerable Presocratic had made his prediction. The problem, however, is that this putative building-block—out of which all things are supposedly compounded—turns out, in truth, not to exist. Its detectable behavior, moreover, proves to be so bizarre that a leading theorist describes it as "a strange kind of physical entity just in the middle between possibility and reality."[2]

Following this problematic breakthrough—which in the wake of the initial euphoria seems to have mystified just about everyone—other quantum particles began to come into view in rapid succession. Meanwhile, in the year 1925, Werner Heisenberg and Erwin Schrödinger, working along very different lines, arrived at what turned out to be inherently one and the same "wave equation,"[3] giving birth to a new physics known as quantum mechanics. What confronts us here, in terms of the Schrödinger formalism, is a so-called wave function ψ, the time-dependence of which is subject to the Schrödinger equation. And in this strange new physics the central mystery manifests even in the simplest cases: the scenario, say, in which ψ is descriptive of a single quantum particle, the point being that the observable properties of the particle—its position or momentum, for instance—have in general no specific value prior to measurement. Instead, there is now a probability distribution which expresses the likelihood of finding the particle in a given region, or its momentum, say, in a given range, whereas in general the quantum particle itself has neither of

2. Werner Heisenberg, *Physics and Philosophy* (Harper & Row, 1958), p. 41.
3. The two formulations, though arrived at from very different points of view and presenting an utterly different appearance, have turned out to be "isomorphic."

these attributes prior to their actual measurement. In consequence quantum particles exhibit weird properties, such as the ability, manifested in the double slit experiment, of seemingly passing through two slits at once. It appears that Heisenberg had good reason to speak of "a strange kind of physical entity"!

The most puzzling thing of all, perhaps, is what happens to a quantum system at the instant of measurement, when that so-called wave function "collapses" to yield a single value of an observable. And what is just as mysterious: if you repeat the experiment a large number of times, the statistical distribution of these measured values is predictable from the wave function itself. The amazing fact is that the physics of these ghost-like and semi-existent particles gives rise to the most accurate predictions known to science—for instance in the calculation of atomic and molecular spectra. At the same time, however, Richard Feynman's "*no one understands*" retains its point.

To which one should add that this reputedly universal incomprehension is by no means due to a lack of effort: it happens namely that ever since the Schrödinger equation was written down, some of the most illustrious physicists have pondered the issue, but with no definitive results. At the core of the impasse stands the conundrum known as the "measurement problem": how to account for that instantaneous "collapse" of the wave function at the moment of measurement; and to the best of my knowledge, the search for a solution continues to this day, with ever more curious results. Take, for example, the well-nigh incomprehensible conception of "superdeterminism," which strikes me as weirder by far than the so-called collapse it is supposed to explain. What I contend is that a *bona fide* resolution of the measurement problem cannot in fact be given in the language of physics for the simple reason that the issue proves to be incurably *ontological*.

Let us remember that physics began with an ontology: the Cartesian dogma of *bifurcation* to be precise—which meanwhile has become ingrained in the physicist's *Weltanschauung* to the point that he is hardly able to recognize that problematic premise as something less than "self-evident." A good reason to ask oneself: what if Descartes was mistaken?

～

The first point to be made concerning Cartesian bifurcation is that the doctrine has no scientific basis: there is absolutely no *experimentum crucis* that could decide the issue. Descartes himself, after conceiving his famous axiom in the seclusion of his garden retreat, felt hard-pressed to justify to himself that a purportedly "external" world exists even so. And it may be of interest to recall that the only argument he could muster to convince himself of this fact was based upon what he termed "the veracity of God." I find it ironic that hard-headed and often enough atheistic scientists should have founded their *Weltanschauung* upon an ontological premise based upon "the veracity of God"!

There were of course—and presumably will always be—philosophers, poets, and mystics of various persuasions who remained unconvinced by the new *Weltanschauung*. Two factors, however, seem eventually to have brought almost everyone amongst the more educated into the fold. In the first place, as time went on and physics continued to progress, a corresponding display of technological marvels kept exposing the public at large to an ever more fantastic array of *"signs and wonders"* that could indeed *"persuade"* the vast majority of the scientistic credo. Add to this the fact that, in course of time, the educational system—literally from grade school up—was in effect taken over by the indoctrinated, now turned faithful promoters of scientism—and it is a wonder that any dissidents are left at all.

～

The deeper one reflects upon the issue, the more evident it becomes that the very possibility of empirical science hinges upon the premise that *we do perceive the external world.* And this applies, above all, to perception of the visual kind. The very first thing Democritus felt obliged to deny, let us recall, was indeed *color*; and when, after two millennia, the Democritean atomism came to be revived, there at the outset was Galileo declaring *color* to be, once again, a "secondary attribute."

I am not minimizing the epistemological difficulty posed by a realist view of visual perception: like every truly fundamental problem of authentic philosophy, it is hard. Error, it seems, is generally easier than the recognition of truth. Yet let us note that the bifurcationist understanding of visual perception condemns us in point of fact to a chronic state of schizophrenia: while "normally" the grass is evidently green, in moments of bifurcationist orthodoxy it instantly turns colorless! To rectify our contemporary scientistic *Weltanschauung*, I say, what is above all called for is a realist understanding of visual perception.

And so, when I set out—in the 1990's—to resolve the measurement problem, I considered it my first task to reestablish the premodern realism regarding visual perception. I elected to do so by means of an inherently phenomenological approach, inspired fundamentally by Edmund Husserl, the mathematician-turned-philosopher of the early twentieth century. The first chapter of *The Quantum Enigma* seeks thus to justify a realist interpretation of visual perception based ultimately upon phenomenological principles. Here is what I said by way of introduction:

Prior to science, prior to philosophy, prior to every ratiocinative inquiry, the world exists and is known in part. It exists not necessarily in the specific sense in which certain scientists or philosophers may have imagined that it does or does not, but precisely as something that can and must on occasion present itself to our inspection. It must so present itself, moreover, by a kind of logical necessity, for it belongs to the very conception of a world to be partially known—even as it belongs to the nature of a circle to enclose some region of the plane. Or to put it another way: if the world were *not* known in part, it would *ipso facto* cease to be the world—"our" world, in any case. Thus, in a sense—which can however be easily misconstrued!—the world exists "for us"; it is there "for our inspection," as I have said.

Now that inspection, to be sure, is accomplished by way of our senses, by way of perception; only it is to be understood from the start that perception is not sensation, pure and simple, which is to say that it is not just a passive reception of images, or an act bereft of human intelligence. But regardless of how the act is consummated, the fact remains that we do perceive the things that surround us: circumstances permitting, we can see, touch, hear, taste and smell them, as everyone knows full well.[4]

On this basis, then, I argued for a realist understanding of *visual* perception, in particular. And so I concluded that Cartesian bifurcation constitutes an ontological fallacy: that the grass is green after all—without bifurcationist interruption—and that, in consequence, Democritus was in truth *dead wrong*.

I was thus able to proceed with my inquiry into the quantum measurement problem, and to arrive in fact quite naturally at a solution. With an added benefit, one might say: it had namely become clear that the inherent stumbling block is not a question of physics at all, but rather of ontology. And given that scientists are generally unschooled in that discipline, the physicist has in all likelihood imbibed the tenet of Cartesian bifurcation unknowingly: it is implicit, after all, in just about everything he has learned in the university and presupposes routinely in his own scientific cogitations.

Getting back to the phenomenological argument: at the very least it shows that the subjectivist interpretation of visual perception is scientifically unfounded and indeed unprovable. And what struck me as I continued to reflect upon the measurement problem was the ease with which that famous conundrum is resolved the instant the Cartesian premise is jettisoned.

4. Sophia Perennis/Angelico Press (2005), p. 9.

~

Before we turn to the physics of measurement, I would like to say a few more words on the subject of visual perception. Some years following the publication of *The Quantum Enigma* it came to my attention that a realist theory of visual perception had actually been proposed—on empirical grounds, no less—by a cognitive psychologist named James J. Gibson. It all began in the early 1940's when, equipped with a fresh Ph.D. from Princeton in cognitive psychology, Gibson received a government grant to devise tests for the selection of military pilots. What stood at issue primarily is the ability to visually determine an aiming point in a state of complex motion—a faculty needed evidently by pilots landing their plane, say on the deck of an aircraft carrier. And to his surprise, Gibson discovered that the information given in the so-called retinal image does not suffice to that end. Now, if such be indeed the case, the argument can be made that this fact alone disqualifies the prevailing "retinal image" theory of visual perception—and that is the conclusion at which the young Gibson did arrive. What followed are decades of hard—and exceedingly brilliant—empirical inquiry into just how we do perceive the external world, out of which emerged what Gibson terms the "*ecological*" theory of visual perception, to emphasize that what is actually perceived is not an image—be it retinal, cortical, or any other kind—but indeed the external *environment* as such. What Gibson argued, on strictly empirical grounds, is that what we perceive visually is not "inside the head," but pertains indeed to the "external" world, the very realm with which the natural sciences are concerned.[5]

5. What I find perhaps most amazing of all is that Gibson was highly regarded within the academic community throughout his career, notwithstanding the fact that his discovery proves lethal to the prevailing and supposedly "scientific" worldview—a fact which might not have been generally recognized. As an introduction to Gibson's "ecological" approach I would suggest his expository treatise: *The Ecological Approach to Visual Perception* (Lawrence Erlbaum Associates, 1986). A summary account of the theory is given in my book *The Vertical Ascent*, op. cit., ch. 5, "Do We Perceive the Corporeal World?"

To my mind, it would be difficult to overestimate the significance of Gibson's discovery, which I strongly believe to be scientifically well-founded. It seems that by reinstating "color" as a
primary attribute, he has disqualified the Democritean atomism
at its root: remember that in the very words of the Presocratic,
"color" is the first thing that *must go*!

To be sure, the Democritean atomism was soon enough
unmasked as a sophism by Plato and the succeeding mainstream
of ancient philosophy; but it appears to me that Gibson may be
the first to discredit Democritean atomism on rigorous *empirical* grounds. He is to be seen as an empirical scientist: a hard-
headed specimen, in fact, who tolerates no humbug. And let me
add that his writings tend to be models of clarity—a fact which
becomes increasingly evident the better we understand him.

It is true that his works could benefit from ontological clari-
fication in basically Aristotelian terms: the concept of *form*, for
instance, is in a way implicit in his "ecological" theory. It is, after
all, what connects the percipient to the external entity—to the
redness, say, in the rose. It might make a marvelous doctoral dis-
sertation—be it in cognitive psychology or philosophy—to interpret Gibson's theory of visual perception in Aristotelian terms.

The first step towards an ontological comprehension of physics
consists then in the rediscovery of what we term the *corporeal*
domain, which is basically the *perceptible*, beginning with the
visual. It is neither necessary nor feasible—initially, at least—to
speculate about the other four senses, nor to worry about "bor-
der cases" and the like. To regain a grasp of "the corporeal" it is
needful to substantiate that the grass is indeed "green": and this
should suffice to break the Cartesian spell.

There is thus a categorical distinction to be made between
the world as conceived by the physicist—which we refer to as
the *physical*—and the perceptible world in which we "live and
have our being," which we distinguish ontologically from the
physical as the *corporeal*.

2

THE MEASUREMENT
QUANDARY

HAVING REJECTED Cartesian bifurcation, we reflect upon what this entails in regard to the measurement problem of quantum theory. And the fact is that, having restored ontological rectitude, the resolution of this conundrum literally "stares us in the face"! This is what we wish now to explain.

In place of the Cartesian dichotomy of so-called *res extensae* versus *res cogitantes*—which has proved to be spurious—another ontological dichotomy has now taken its place: one needs namely to distinguish ontologically between the entities of physics and the entities comprising this ordinary perceptible world. What invalidates the Democritean premise is the fact that *qualities*—beginning, yes, with *color*—have once again been recognized as *bona fide* components of the "external" world. And this has drastic implications for physics, given that—by its very *modus operandi*—it has no eye for *qualities*! A worldview based upon physics is bound, therefore, to exclude the "qualitative dimension" of the cosmos—not because it is not there—but because this science is categorically incapable of grasping that dimension, that aspect of the world.

An ontological distinction is thus to be made between the spatio-temporal world as *perceptible*, and that world *as conceived by the physicist*: the two are by no means the same. No wonder that—from the physicist's point of view—the measurement problem could not be solved: what else can you expect, given that the act of measurement—inasmuch as the result must be *perceptible*—transcends "the world of physics" categorically! What strikes me as perhaps the most amazing fact concerning the quantum measurement quandary is that it has apparently failed to rouse the physics community at large from their "Cartesian slumber."

In keeping with the conclusion reached in chapter 1, what is called for is an ontological distinction between the spatio-temporal world as perceptible and the world as conceived by the physicist: the *corporeal* and the *physical*, we shall say. It appears that Eddington was right: there *are* in truth "two tables."[1] There is namely the one I can see and touch, and the one made of "atoms and the void" which I cannot.

～

To gain maximal clarity we need to recognize that there exists also a function which to every *corporeal* object X assigns the associated *physical* object SX. And it seems that physicists pass back and forth between X and SX without recognizing the ontological discrepancy separating the two: they tend namely to identify X and SX, an error which renders the quantum measurement problem insoluble.

What confronts us in the act of measurement is a transition between two distinct ontological domains: from the physical to the corporeal that is. Obviously so, given that the quantum system is physical and the measuring instrument corporeal—which evidently it needs to be to render the result of the measurement perceptible. There is then a transition from the physical to the corporeal—from an SX to X—which is however invisible to

---

1. The fact that someone distinguishes these "two tables" does not imply that he shares the ontology we propose, the point being that the latter is based upon the ontological discernment argued in chapter 1. Everything hinges upon a realist understanding of perception, beginning with the visual. Regarding Eddington's "two table" commentary, see *The Nature of the Physical World* (New York: Macmillan, 1928), pp. ix ff.

physics as such, for two reasons: ontological and etiological. This transition is, first of all, inconceivable to the physicist ontologically, because *qua* physicist he cannot conceive of the corporeal domain; and it is inconceivable to him etiologically because a transition between two ontological domains can only occur *instantaneously*, whereas the causation known to physics transpires in time. What is called for, consequently, is a mode of causation that *does not transpire in time*: *vertical* causality, I call it.

Before going farther I would like to note—as already mentioned in the Preface—that three years after I posited vertical causation, its existence was confirmed by a mathematician named William Dembski in the form of a theorem which has since become rather well known under the designation "intelligent design." Dembski proved that there is one thing horizontal causation[2] cannot produce: i.e., what he terms *complex specified information* or CSI. We need not concern ourselves with the specifics of Dembski's proof, which is information-theoretic; it happens that the result is inherently ontological, and can be argued quite simply on purely ontological grounds.[3] What concerns us is the fact that horizontal causality—with which physics is exclusively concerned—is complemented by an altogether different kind, what I refer to as *vertical* causation, which is able to effect transitions inconceivable to physics as such.

Getting back to the measurement problem, we have come to see that its resolution cannot be achieved on the basis of physics, but requires two principles extraneous to physical science: the concept of *corporeal* as opposed to *physical* entities, and that of *vertical* as distinguished from *horizontal* causation. With these metaphysical categories in place, *the resolution of the measurement conundrum proves to be virtually "instantaneous."*

What is more, physics ceases thus to stand alone—in self-proclaimed isolation from the accumulated wisdom of

2. By horizontal causation I mean the causation underlying physics, which operates in time by way of a transmission through space.

3. See Part II, ch. 1.

mankind—but finds its place in the hierarchy of human know-ing. And this entails that physical science can no longer be invoked to reduce all that transcends its purview to the status of a so-called "pre-scientific superstition," a practice which since about the seventeenth century has dangerously impoverished Western civilization and threatens to usher in a new barbarism.

But these are of course digressions—historical and cul-tural—from which I will henceforth desist.[4] Our focus is now upon the measurement problem; and I venture to say that with the distinction between the *corporeal* and the *physical*, together with *vertical* versus *horizontal* causation, that problem has been solved. It has been solved moreover without the aid of phys-ics, for the simple reason that *physics as such has nothing to say in regard to this question*: its equations simply don't reach that far. The crux of the matter resides in the fact that the physi-cal realm—the "world" in which these equations do cut ice—is limited, and that the act of measurement cannot be consum-mated within that restricted domain.[5]

Like it or not, that act terminates in a world from which "color," for instance, has *not* been banned. We will, of course, be told instantly that physics comprehends color inasmuch as the latter "reduces" to frequency or wavelength: yes, from the stand-point of physics it most certainly does. Yet the fact remains that *frequency* belongs to SX, whereas *color* resides in X itself; and there is literally "a world of difference" between the two—a gap physics cannot span. This science came to birth in the somewhat imaginary universe of *res extensae*, which in a way our physics seems still to inhabit. Yet—sooner or later—exit it must; which is to say that the time to jettison the Cartesian premise may be at hand.

---

4. I have dealt with these wider and indeed "cultural" issues at con-siderable length, beginning with my first book: *Cosmos and Transcendence* (reprinted by Philos-Sophia Initiative Foundation in 2021).

5. For a fuller treatment of the measurement problem I refer to *Physics and Vertical Causation* (Angelico Press, 2019), ch. 3.

There are three fundamental ontological conceptions, we maintain, which it is incumbent for physicists to recognize. And the first two have already been specified: they are the corporeal as opposed to the physical, and vertical causation as distinguished from horizontal. We propose, in the next chapter, to introduce the third.

3

PHYSICS AND
IRREDUCIBLE WHOLENESS

WITH THE PUBLICATION of *The Quantum Enigma* in 1995, one could understand physics as the science of the physical, and the act of measurement as a transition from the physical to the corporeal effected by vertical causation. Well and good.

Meanwhile however a third ontological factor has come into view, the implications of which for physics as such we have sought to discern. What needs in consequence to be added breaks into three steps, the first being of course to define that additional third principle; the second is to reflect upon its ontological implications, which can perhaps be described as a partial "rediscovery" of the Platonist metaphysics; and the third is to investigate what this implies regarding physics at large, both classical and quantum-theoretic.

❧

The ontological concept in question can be fitly denominated *irreducible wholeness*: a wholeness, that is, which does not reduce to a sum of parts. And let us note immediately that such an IW is incomprehensible to physics *per se*, for the very simple reason that its *modus operandi* hinges precisely upon the reduction of wholes to a sum of their spatio-temporal parts. Or to put it another way: what renders IW incomprehensible to physics is that the latter constitutes a science of the *quantitative*.

The first point I wish to make is that this idea of *irreducible* wholeness goes hand in hand with that of *vertical* causation, which in fact "admits no parts" at all. It can therefore be seen—with the "eye of *intellect*" as the ancients would say—that, on the one hand, *it takes VC to produce an IW*, and conversely, that *it takes an IW to exert VC*. And let us not fail to note that the first of these recognitions generalizes Dembski's so-called "intelligent design" theorem,[1] which affirms—as we have noted before—that no deterministic, random, or stochastic process can give rise to "complex specified information," a result which can indeed be viewed as a highly special case of the aforesaid ontological affirmation.[2]

I would point out, moreover, that both VC and IW are inherently antipodal to the causality and wholeness, respectively, as conceived in physics *per se*. This is self-evident in the case of VC inasmuch as an "instantaneous" effect cannot be accounted for in terms of a causality operative in time. That antipodal quality is likewise apparent, however, for IW inasmuch as the *modus operandi* of physics is based upon a reduction of wholes to a sum of their "atomistic" parts, defined ultimately by the familiar spatio-temporal quadruple (x_1, x_2, x_3, t).

This brings us to the very point upon which—as we shall presently see—the authentic ontology of physics rests: which is that a corporeal entity X does not reduce to any such decomposition, to any "sum of parts": i.e., that *a corporeal entity X is an IW*.

∿

What renders a corporeal entity thus irreducible is the ontological fact that, *in its essence, it is not subject to the bounds of space and time*. And this notion of irreducible wholeness leads quite naturally to the Platonist ontology represented by the tripartite

1. *The Design Inference* (Cambridge University Press, 1998).
2. On this issue I refer to Part II, ch. 1.

cosmos, as we have come to conceive of it.[3] In this ontology the integral cosmos divides hierarchically into three domains, ranging from the *corporeal*—subject to the bounds of both space and time—to what may traditionally be termed the *aeviternal*, which is subject to neither and must consequently be the ontological stratum from which both VC and IW are ultimately derived. Dante refers to it poetically as *"the pivot around which the primordial wheel revolves,"* a metaphor in which that *"primordial wheel"* refers evidently to *time*, which proves in fact to be definitive of an ontological stratum referred to traditionally as the *intermediary* domain, of which the corporeal is the termination. The integral cosmos may therefore be represented in iconographic mode as a circle in which the center represents the aeviternal realm, the interior the intermediary, and the circumference the corporeal domain.[4]

A major observation is to be made at this juncture. The fact is that the existence of an ontological stratum subject to the bound of time alone invalidates Einsteinian relativity at a single stroke. Or to put it another way, the validity of relativistic physics would disqualify the tripartite cosmology which I regard to be both factual and authentically Platonist. Suffice it to say that I have elsewhere treated this issue at length.[5] Based upon an abundance of empirical evidence as well as on ontological grounds, I have become fully persuaded that *there is no four-dimensional space-time*, and that a physics based upon space-time symmetries is simply erroneous. Two facts should perhaps be mentioned to dispel the initial impression that such a "contra Einstein" position could not possibly be true: first, that the enormity of the light velocity—which is

---

3. See ch. 2 of *The Vertical Ascent*, op. cit., entitled "The Tripartite Wholeness."

4. For a detailed discussion of this representation and what it entails, I refer to *Physics and Vertical Causation*, op. cit., ch. 8. Regarding the intermediary domain, see also Part II, ch. 4.

5. *Physics and Vertical Causation*, op. cit., ch. 5.

approximately 300,000 kilometers per second—renders the discrepancy between Einsteinian and non-Einsteinian predictions almost unmeasurably small; and secondly, that the famous "atom bomb" formula $E = mc^2$ is equally valid on a nonrelativistic basis.

At first glance it may seem that relegating IW to the so-called aeviternal realm is to place it "out of reach," as it were, for us mortals; but as a matter of fact, just the opposite holds true. That "*pivot*," namely, around which "*the primordial wheel*" is said to "*revolve*," turns out to be the *nunc stans*—the ever-present "now"—of which Dante goes on to say: "*There every where and every when are focused... Heaven and all Nature hang from that point.*"[6]

I would emphasize that—so far from constituting "mere poetry"—this is sound Platonist ontology as well, which in fact provides the indispensable key to an *ontological* understanding of physics. The catch is that Dante speaks on a level above the normal: the level of *intellect* as opposed to *reason*—a distinction which apparently did not survive the Enlightenment. Yet it is both real and absolutely essential—and actually not that difficult to comprehend. The notion of intellect can perhaps be most readily explained in the context of music: in terms of the difference, that is, between a sequence of tones—which anyone whose hearing is not impaired can discern—and a *melody*, which is something else entirely, and can be discerned only on the plane of intellect. And this explains a decisive point made by Wolfgang Amadeus Mozart when he confided to a friend that "an entire symphony comes into my mind 'all at once': in a single instant." That "single instant" is the *nunc stans* which all of us "enter" so to speak whenever we activate the faculty known to the ancient philosophers as *intellectus* in contrast to *ratio*. What enables us to do so—and this is another perennial recognition roundly forgotten in modern times—is the fact that *man himself is tripartite*, consisting, in traditional parlance, of *corpus*,

6. *Paradiso* xxix, 12; and xxviii, 41. The initial quote was from xiii, 11.

anima, and *spiritus*. The integral human being is thus actually a *microcosm*—failing which we could not grasp the cosmos at large, what the ancients termed the *macrocosm*.

∽

The fact is that *every corporeal object constitutes an irreducible wholeness* which is in truth its very *being*: "*ens et unum convertuntur*"—being and oneness are equivalent—declares a Scholastic dictum. The aeviternal realm is inherently the cosmos in its integrality "before" it is fragmented into innumerable "bits and pieces" through the imposition of temporal and spatial bounds. Yet despite this partition into spatio-temporal fragments, every corporeal entity X remains "undivided" in its very being. It therefore constitutes—here and now—an irreducible wholeness, which moreover survives its disappearance from the spatio-temporal realm even as it antecedes its spatio-temporal manifestation.

Consider a very simple corporeal entity: a pebble, let us say. The point is that this entity—this pebble—is yet incomparably *more* than the physicist takes it to be: more in fact than, *qua* physicist, he *can* take it to be. For whereas he is bound, by the *modus operandi* of physics, to regard this entity as the sum of its spatio-temporal parts, *qua* IW it actually pertains—here and now—to the aeviternity of the supra-temporal order. Evidently so, given that—*qua* irreducible wholeness—it never ceases to reside in that supernal realm! In the final count, it is this transcendence of the spatio-temporal that bestows upon that innocuous-seeming entity—that pebble—its corporeal identity. And let us observe, parenthetically, that from a Platonist perspective the notion of an "evolutionist origin" proves to be incongruous to the point of self-contradiction.

∽

I would note that the Pythagorean and Platonist traditions

bear witness to the presence of IW even in the quantitative order, a matter I have touched upon elsewhere.[7] One can understand this by way of music, once again. Think of the *octave* say, and of the even more mysterious *major/minor* dichotomy—that a single half-tone can take you from the active "solar" world to the introspective "lunar," which anyone who has heard this himself will understand very well—and you will realize that what confronts us here are indeed matters of *irreducible wholeness.*

The truth is that IW does have a place in pure mathematics, whether the contemporary authorities recognize this or not. There has been a concerted effort, namely, to "atomize" mathematics as such: to reduce the discipline inherently to set theory. And it was René Descartes who took the first decisive step in that regard by "arithmetizing" geometry through the introduction of what, to this day, is termed a Cartesian coordinate system.

The final step in this reduction was taken by Whitehead and Russell in their *Principia Mathematica*, which claims to reduce mathematics to set theory *without loss or remnant.* I do not deny that their three-volume masterpiece has enabled prodigies of mathematical acrobatics, such as that of the justly famous Gödel theorems; my point is simply that this achievement was not without cost. Something of incalculable import has been lost to the mathematical sciences: *irreducible wholeness,* that is!

And I would add that *Principia Mathematica*, though read by only a stalwart few, has even so had a powerful impact upon Western civilization: in outlawing IW, as it were, it has promoted a leveling down of Occidental culture, a drastic flattening thereof *in toto.* I surmise that the nihilistic nominalism of the point set paradigm has eventually filtered down to the public at large as the last word of Science, and in so doing has contributed decisively to the "enlightenment" of millions. One

7. See for instance "Irreducible Wholeness and Dembski's Theorem" (Part II, ch. 1).

might say that it has disenchanted the universe, deprived it, in the eyes of the "enlightened ones," of the authentic mystery residing in every natural thing—down to the humblest. And so it has in effect dehumanized a significant portion of mankind by rendering it profane.

The fact is, nonetheless, that the *domain of quantity does not exclude irreducible wholeness*; if such were the case, not only could there be no music or traditional arts, *but there could be no physics as well*: that is what we propose now to prove.

∽

What we have to work with are ontological connections between irreducible wholeness and vertical causation which prove to be so basic as to be "seeable" in an intellective sense. We shall refer to these as ontological "axioms," which in a way is precisely what they are. There are three such tenets that will concern us, the first being the previously mentioned "ontological" Dembski theorem. The "specification" part of the "complex specified information" or CSI requirement identifies the "target" of a given process—which can be deterministic, random, or stochastic—as an IW, whereas the "complexity" requirement is there to rule out an "accidental" hit. It is thus the information-theoretic setting that accounts for the complexity requirement in Dembski's theorem, whereas specification serves to identify the target as an IW. Neither of these conditions, needless to say, enters into the ontological version, which affirms quite simply that

(I) *It takes VC to produce IW.*

There are two more such ontological "axioms":

(II) *The causation emanating from an IW is vertical.*

(III) *The product of VC is an IW.*

And this brings us to our fundamental contention: i.e., that a *corporeal entity X is an IW*. We need however to specify more precisely what we mean by a "corporeal entity": it will not do to admit what is technically called a *mixture*,[8] which evidently is not an irreducible wholeness. We will henceforth restrict the term "corporeal entity" to the case where X is a substance: is endowed, in other words, with a substantial form. The proof that X constitutes an irreducible wholeness is immediate:

> Inasmuch as the substantial form is an IW, it follows by (II) that it acts via VC, and by (III) that it is productive of an IW, as was to be shown.

∽

This brings us to SX, what in *The Quantum Enigma* we defined to be "X as conceived by the physicist," which in a way it is. Or better said, perhaps, SX is "X conceived in bifurcationist terms as a *res extensa*"—which in the given historical circumstances it not only *is*, but in a sense *has to be*. Yet, from an ontological point of view it must be something more, which is to say that what interests us now is not SX as an intentional object, but as a *being* of some kind. We shall consequently look at X—neither from a positivistic nor from a phenomenological—but from an authentically ontological point of view, to understand *how SX is related to that "more than physical" X*.

Since X is not a *set* but a *substance*, SX cannot be simply a subset. What it must in fact be is a *part*: a part of a whole. But this needs now to be understood in an appropriate ontological sense, the point being that, inasmuch as X is an IW, it is not just a set, which entails that a part of X cannot be a subset either. It appears thus that SX cannot be a part in a set-theoretic sense, but must—*qua* part—"*participate*" somehow in the wholeness of X.

8. In Scholastic parlance, a mixture is an ensemble of substances.

It is easy to see that such is indeed the case in the biosphere: an organic part of a living organism—a limb, say, or an organ— participates evidently in the wholeness of the organism. Or take a work of art, say a symphony or a sonnet: here too every distinguishable part participates in the "oneness" of the whole. And is this not why a connoisseur can, from a single line of that sonnet, recognize that it is Shakespeare, for instance? If the whole is more than the sum of its parts, a part too is more than simply a part. It follows that *a part of an irreducible whole participates in its wholeness.*

What might render this hard to comprehend is the tyranny of the set-theory paradigm to which we have referred, and which appears to have imposed itself upon present-day civilization. But the cosmos at large, happily, is not based upon that paradigm, otherwise there could be no living organisms, no sonnets, and for that matter, no physical science either—that is what we must now explain.

In formal ontological terms, what stands at issue is another "seeable" fact, analogous to our three ontological "axioms," which can indeed be proved with their aid. The argument can run like this: since X is an IW and SX is a part of X, X acts upon SX, which implies by (II) and (III), that SX is an IW.

The gist of the matter is that inasmuch as X constitutes an irreducible wholeness, one cannot actually "separate" SX from X. At the time I wrote *The Quantum Enigma*, I sensed the ontological mystery of X—the fact that it is not reducible to a sum of parts—but held what might be termed a "profane" view of SX. Today, on the other hand—more than two decades later—I realize that the ontological mystery of X renders SX mysterious as well: it does so, namely, by *rendering SX itself an IW.* And this "mystery" proves to be decisive: its implications impact all the domains of physics, from the classical to the quantum-theoretic, where in fact that irreducible wholeness of SX is the reason why "no one understands" quantum theory.

Meanwhile that ontological verity has startling implications for subcorporeal physics—i.e., the physics of an SX—which have but recently been discovered by physicists, to their consternation. It follows namely that the IW of SX, which we have just established on ontological grounds, has manifested in terms of a phenomenon pertaining to subcorporeal physics which in recent years has captured the interest of physicists. Barbara Drossel and George Ellis, for instance, motivated by the measurement problem, have taken a new look at the physics of a measuring instrument;[9] and what they found is that SX[10] contains a so-called heat bath which is quantum-mechanical locally and thermodynamic globally. That physical structure proves therefore to constitute an irreducible wholeness in compliance with the aforesaid ontological recognition; and let us add that this in turn implies, by (I), that the heat bath cannot be the product of horizontal causation, which is to say that its origin is incomprehensible to physics as such. In a word, the "mystery" of the heat bath has been resolved on the only basis it *can* be: i.e., on *ontological* ground.[11]

∾

Following upon these foundational considerations pertaining to *subcorporeal* physics, it behooves us to consider its complement: the physics of what, in *The Quantum Enigma*, we termed the *transcorporeal*. This is the physics of entities which are not the SX of a corporeal X: the physics of physical entities, thus, standing supposedly on their own. The definitive fact is this: whereas, in the ontological descent to the subcorporeal, what

9. "Contextual Wavefunction collapse: an integrated theory of quantum measurement," *New Journal of Physics* 20, 113025 (2018).

10. Admittedly a measuring instrument is, strictly speaking, a "mixture." The decisive transition from SX to X takes place, however, in a component which constitutes a substance.

11. For a detailed treatment of subcorporeal physics and the heat bath phenomenon, see Part II, ch. 2.

is lost are the qualities, in the transcorporeal one thing more is evidently forfeited: *irreducible wholeness* namely, which in the subcorporeal, as we have seen, comes to SX directly from X. It may consequently be supposed that, in the transcorporeal *there is no* IW—unless, of course, it is brought in by some other means.

To grasp what this entails, we need to recall the basic onto-logical fact expressed in the dictum "*ens et unum convertuntur*": this means that without irreducible wholeness there is no *being* as well. And this, I take it, is vital to the comprehension of quantum theory, which is the physics of entities void of IW, and therefore of *being*. Is it any wonder, then, that "no one understands"?

It is not surprising, first of all, that the very formalism of quantum theory should be categorically different from that of classical physics. The classical mechanics of a system of par-ticles, for example, is concerned with their masses, positions, velocities and accelerations—in short, their measurable attri-butes—but does not "mathematicise" the particles themselves. Contrast this now with the quantum mechanics of Heisenberg and Schrödinger: the astounding fact—utterly incomprehensi-ble from a "classical" point of view—is that, in the new physics, *the state of a physical system is itself conceived as an element of a vec-tor space*. A state can namely be multiplied by a complex num-ber, and two states can be added. One can thus, for example, add the state of the system in which a given so-called particle is, say, in region A, to a state in which it is in some other region B, as widely separated from A as you wish. And *voilà*: you have now a bilocating particle!

How, then, is one to interpret such prodigies: does it mean that quantum particles do actually "multilocate"? The reso-lution of this conundrum resides in the recognition that, in the transcorporeal, there *are no* "real" or "actual" particles: on ontological grounds, as we have noted, *there cannot be*. In the quantum realm—which *is* the transcorporeal—we are no lon-ger dealing with existent things. As Heisenberg recognized so clearly from the outset, the new physics has to do with mere *potentiae*: weighted possibilities, one might say; and so too, what it normally yields are *probabilities.*

The ontologist, on the other hand, knows from the start that transcorporeal physics must also derive its efficacy from that *"pivot around which the primordial wheel revolves"*—for the very simple reason that all things cosmic do. And this brings us back to Werner Heisenberg, who once again hands us the key to the enigma when he observes:

> Science [meaning physics] no longer stands before Nature as an onlooker, but recognizes itself as part of this interplay between Man and Nature.[12]

Now this is exactly the point! It means, in plain words, that *the quantum realm originates as the response of Nature to questions posed by the physicist himself.* How does the physicist do so? Clearly, it can only be done by means of *corporeal* instruments. And how then does Nature respond? But this happens to be precisely the enigma of quantum measurement, which is resolved by the recognition that this too is accomplished by *corporeal* instruments acting by way of *vertical* causation.

It turns out, thus, that *the quantum realm derives likewise in a sense from the corporeal*—but in a categorically different way than the classical! In subcorporeal physics we are dealing with entities SX which receive their IW—their *being*—directly from a corresponding corporeal substance X. In transcorporeal physics, on the other hand—that is to say, in the quantum realm— the connection derives from the strategies of the experimental physicist, consisting of a two-stage process: there is an "asking of the question," followed by a "receiving of the response."

A vital fact presents itself at this point: inasmuch as the physicist's "questions" are posed via corporeal instruments—which,

---

12. *Das Naturbild der heutigen Physik* (Hamburg: Rowohlt, 1955), p. 21.

as we have seen, are made up of irreducible wholes—it follows by (II) and (III) that *IW is thereby transmitted into the transcorporeal realm*. And this may indeed be crucial, given that the transcorporeal as such *has no IW of its own*. It may be this transmission of IW into that "beingless" realm that enables a new kind of physics: the kind in which Man is no longer just an "onlooker," as Heisenberg perspicaciously noted, but assists in a way to produce what he observes.

Being may thus be transmitted into the transcorporeal—not to stay there—but to return: the first stage of the process is there for the sake of the second. And the entire process is powered ultimately from that central Point of which Dante speaks: for *being* as such, together with all causality, originates from that aeviternal "*pivot*" around which "*the primordial wheel revolves*." There are no exceptions to this rule: the "chain of *being*" descends centrifugally from thence to the furthest extremities, which can presumably be none other than the transcorporeal, the realm of quantum theory.

One might add—in light of these ontological reflections—that whenever physicists are forced to leave the comparative *terra firma* of the merely technical to explain to the public at large what "the new physics" is all about, they can hardly do other than drift into the realm of science fiction: the case of Werner Heisenberg—the son of a classicist—is indeed exceptional.

~

Complementary in a way to Heisenberg, mention should be made of Sir Arthur Eddington, whose forte was *epistemological*. He appears to be the physicist who understood—more profoundly, perhaps, than anyone else—how much of what quantum physics "discovers" is in truth the rigorous consequence of its strategies; and in so doing he corroborates, in a way, what Heisenberg has to say regarding the "participatory" nature of the new physics.

Eddington's epistemological reflections, moreover, have led, often enough, to mathematical calculations: the determination, for instance—with zero empirical input—of the famed "fine structure constant," which he finds to be 1/137—a result that strikes me as being far too close to the best measured values to be written off precipitously as a mere coincidence. The fact is that, by way of his epistemological approach to the philosophy of physics—which he referred to as "selective subjectivism"—Eddington arrived at a realization which accords fundamentally with the position of Heisenberg: both physicists concluded that *"the mathematics is not there till we put it there,"* as Eddington states.[13]

And from an ontological perspective, it could not be otherwise: strictly speaking, in the transcorporeal there is in fact no "there" until what Heisenberg terms "this interplay between Man and Nature" has been initiated. Entities void of irreducible wholeness, after all, do not simply stand in place like Mount Everest: to be precise, they are not actually "discovered," but are in a sense "made." The ineluctable fact, namely, is that the transcorporeal domain derives whatever reality it appears to have from the corporeal, and this takes place—"artificially" if you will—by the strategies of the physicist, as both Heisenberg and Eddington affirm, howbeit from very different points of vantage.

Shifting from what quantum physics *is* to what it can *do*, I would note that it has in fact given rise to the most accurate predictions of physical science ever made: in countless experiments of unprecedented precision it has emerged triumphant. It may well be this spectacular prowess that has persuaded so many physicists to hypostatize the quantum realm, elevating it to the status of the prime reality—which in truth it is not, and cannot be.

---

13. *The Philosophy of Physical Science* (Cambridge University Press, 1949), p. 137.

~

Historically speaking, transcorporeal physics originated under the aegis of logical positivism and kindred philosophical schools. It was in the early decades of the twentieth century that leading physicists became acutely aware of the distinction between "measurable" and "unmeasurable" quantities, and came to be seized by a compulsion to eliminate all quantities of the latter kind. As Eddington points out:

> Previously scientists professed profound respect for the "hard facts of observation"; but it had not occurred to them to ascertain what they were.[14]

The time did come, however, when in certain groups—such as the prestigious Vienna Circle—the "hunting down" of unobservables became the laudable scientific thing to do. For Albert Einstein it may have been "simultaneity of distant events," whereas for others it was, say, the "position-and-momentum" of a particle. In either case the unobservable had to go—regardless of the cost. And at least when it comes to the physics of "the very small," the strategy proved to be spectacularly successful: it paved the way to the discovery of quantum physics.

Yet admirable as this passion for perfect positivistic rigor may be, the cost was greater than anyone imagined. Along with the offending unobservables, something was jettisoned one might have wished to keep: *being*, namely. And it was Eddington who promptly noted the loss; commenting on the transition from classical to quantum physics, he observed that "*the concept of substance has disappeared from fundamental physics.*"[15] What has actually disappeared from fundamental physics, however, is not only "the concept of substance," but *substance itself.*

14. Ibid., p. 32.
15. Ibid., p. 110.

Whatever his train of reasoning may have been, Eddington arrived at the conclusion that what you "catch in your net" is predictable from the mathematical structure of the net itself—not, to be sure, in all its details, but in its primary laws and basic dimensionless constants. And in a fascinating chapter entitled "Discovery or Manufacture?"[16] he seems to strike the target when he points out that "substance might have offered some resistance to the observer's interference." Arguing the side of "manufacture" as opposed to "discovery," he makes a powerful case when he explains that "under the cover of the term 'good' observation, the bed of Procrustes is artfully concealed." And this, I surmise, must indeed be a major part of the story of how the quantum world is actually made: for "*made*" it surely is—in keeping with Heisenberg's observation that Man is part of this "interplay with Nature."

Approaching the matter from an ontological point of view, we too have concluded that the transcorporeal realm does not "stand alone": that in fact it *cannot*. True to the scenario articulated by Heisenberg, it is by means of *vertical* causation emanating from *corporeal* instruments that the physicist asks the critical questions, and it is by the same means that he receives the definitive response. The point is that, *apart from this interplay*, there *is no* quantum world: as we have come to see on traditional ontological grounds, there absolutely *cannot be*. Whatever reality, namely, is to be found anywhere within the cosmos must ultimately derive from its aeviternal "*pivot*"; and in the physical, this means by way of the corporeal. And whereas in the subcorporeal realm this transmission of VC is direct, in the transcorporeal it occurs evidently by way of Heisenberg's "interplay" involving corporeal entities *plus the physicist himself*, who—in addition to VC emanating from corporeal instruments—*contributes also a VC input of his own.*

Admittedly our account of quantum physics deviates radically from the narrative recited dogmatically in universities

16. Ibid., pp. 106 ff.

around the world as "scientifically proven" fact. Based upon ontological principles that might be labeled Platonist, we have arrived at a view of quantum theory which accords fundamentally with the position of both Heisenberg and Eddington. It has been aptly summarized by the British savant in the memorable words: "*We have discovered a strange foot-print on the shores of the unknown... And Lo! it is our own.*"[17]

∿

According to a widespread and supposedly well-informed opinion, quantum theory is inherently the physics of the microworld. What confronts us here is basically the Democritean claim that cosmic reality derives from minute "atoms" out of which all things "macroscopic" are said to be compounded. And in this optic classical physics reduces evidently to that of the *macroworld*: the realm of entities distinguished by the fact that they are several orders of magnitudes larger than the "atoms" out of which they are supposedly composed. The difference between the microworld of quantum theory and the macroworld of classical physics is thus reduced to a matter of *scale*: whether one looks at the universe "in the small" or "in the large." One assumes that the physics of the macroworld has to do with statistical averages over vast ensembles of microphysical entities: that it deals thus with a kind of "secondary reality," whilst the primary is that of the microworld, the realm of "atoms and the void"—just as Democritus opined in the 5th century BC.

For the sake of maximal clarity—and at the risk of being repetitious—I would like now to show, in light of the preceding considerations, that this Democritean *Weltanschauung* is not simply wrong, but actually "upside down."

Certainly we do not deny that—in a duly qualified and

17. *Space, Time and Gravitation* (Cambridge University Press, 1921), p. 201.

greatly attenuated sense—there *are* "fundamental particles"; what we do deny categorically is that these particles constitute the reality out of which the entities of classical physics are compounded—which is another matter altogether. My point is that the physical entities which make up the macroworld are in truth subcorporeal, which is to say that they derive their reality—not from the microworld—but from the opposite direction: not thus "from below," but "from above," precisely. And that supernal reality, moreover, is not a matter of "particles," but is something, rather, of which physics *per se* cannot so much as conceive: it is namely an *irreducible wholeness*.

And this brings us back to an absolutely fundamental question: what exactly is the ontological status of quantum particles? The pivotal fact has already been noted: instead of bestowing reality upon the macroworld, these putative particles actually receive whatever reality they possess by way of interaction with corporeal objects. One sees thus, on strictly ontological grounds, that Heisenberg had it exactly right when he observed that quantum particles are something "midway between being and nonbeing." The subcorporeal realm cannot "emerge" from the transcorporeal simply through an aggregation of quantum particles, followed by an averaging over the resultant aggregates, as one tends nowadays to imagine: one forgets that, in the transcorporeal realm, there *are no actual particles*! *Real* particles—if there be such a thing—do not "emerge" from *potentiae* as if by magic: what is called for *is an active or efficient cause*.

The matter is in truth exceedingly simple: whatsoever can be produced through an aggregation of parts is *ipso facto* reducible to these parts, and *cannot therefore constitute an IW*. And let us not fail to note that this ontological recognition suffices to *refute atomism* in all of its multifarious forms: not a speck of dust can "emerge" out of a microworld consisting of mere *potentiae*! Physicists could have saved themselves the trouble of inventing "inflation," "dark matter," and a vast conglomerate of other such postulated constructs, not to speak of the burgeoning ensemble of ever more exotic particles! Fascinating as all this may be to the specialist, *it cannot take us a single step beyond the realm of potentiae.* There is simply no getting around the ontological fact that the entities of classical physics derive their being—not from the quantum realm—but from the *corporeal.* And this absolutely basic recognition renders the transition from quantum theory to classical physics *incomprehensible to the physicist as such.*

To be precise: inasmuch as physics operates by way of horizontal causation, the only wholeness it knows—and *can* know—is the kind that is reducible to its parts. Physics cannot, therefore, conceive of a subcorporeal entity—let alone account for its existence. As Lord Kelvin has so aptly pointed out, physics is indeed "the science of measurement"; and that defines the domain to which it is consequently restricted. The very fact, however, that physics is geared exclusively to deal with the realm of quantity renders it incapable of comprehending other matters—beginning with its own foundations.

Physics as such can tell us neither what a subcorporeal object is, nor where it comes from; and worst of all perhaps, it cannot recognize that inherent incapacity. In a word: outside the quantitative domain, physics can tell us nothing at all. This is not to say, of course, that the physics community is neutral in matters of ontology; it merely affirms that whatever their proffered ontology might be, it can claim no support or validation on the basis of physics *per se.*

It emerges that quantum particles can become actual components of existing entities *only* by being "actualized" through contact with a subcorporeal entity: with the SX, that is, of a corporeal object X. Or to put it another way: to actualize a quantum-theoretic *potentia* requires an act of *vertical* causation, which in the physical realm can only derive from an SX. It must not be thought, however, that the particle itself transitions, as it were, from potency to act. The point is that there are no *actual* quantum particles: neither before nor after an encounter with an SX can there be such a thing! What happens—what alone *can* happen—is that, in consequence of such an encounter, the quantum particle *ceases to be a quantum particle* to become—not an "actual" quantum particle—but a *bona fide* "part" of SX. And let us note that this can only happen by virtue of *irreducible wholeness*: only a wholeness that is "more than the sum of parts" can thus "incorporate" into itself a formerly "foreign" entity. Moreover, that incorporation—that "miracle" if you will—can only be effected by an act of VC emanating from that IW: quantum particles transition from potency to act by ceasing to be—not just quantum particles—but "particles," period.

～

Let me touch now upon some of the major insights quantum mechanics has supplied, which stand solid as a rock. First of all it has proved decisive in the understanding of chemistry, beginning with the classification of the chemical elements. The periodic table of elements, which emerged gradually on a purely empirical basis, began abruptly to make sense, isotopes and all. A vast conglomerate of facts—seemingly "without rhyme or reason"—became, almost instantly, comprehensible the only way, apparently, it could: on the basis of protons, electrons and neutrons, namely! The fact that these so-called particles have been misconceived ontologically does not affect the validity of the resultant theory of the chemical elements in the least, the point being that this theory is physical—or quantitative—as opposed to ontological.

Not only, however, does quantum mechanics bring order into the maze of the chemical elements, but it leads to a supremely accurate theoretical understanding of their observable properties, a breakthrough that constitutes one of the most impressive achievements of contemporary science. Hardly, for example, had Erwin Schrödinger written down his quantum-mechanical "wave equation" than geniuses like Wolfgang Pauli began calculating atomic spectra, which could then be checked against spectroscopic observations. And guess what: the predictions matched the observed values to degrees of accuracy previously unknown in physics.

Let us then not spoil magnificent physics with a spurious ontology for which there is not a shred of evidence, and become open instead to what appears to be the crucial fact:

> *The transcorporeal realm consists of potentiae to be actualized upon contact with a subcorporeal entity SX by means of VC emanating from the IW of X.*[18]

And here, in this absolutely basic recognition, we have the reason, I maintain, why "no one understands quantum theory": it appears, namely, that in the absence of this key ontological insight, *no one can.*

～

As we have come to see, IW invalidates atomism at a single stroke by showing that "atoms" of whatever description do not add up to yield *being*—for the very simple reason that *being* does not reduce to a sum of parts! It matters not whether we conceive of these putative atoms crudely, as tiny "nuggets" of this or that, or with the uncanny sophistication of present-day particle physics: the fact that *being* does not reduce to a sum of parts takes care of all such conceivable scenarios—from Democritus all the way to CERN.

One needs at the same time to realize that—in a suitably restricted sense—the particulate model of the chemical

---

18. To those who might argue that we should have said "the IW of SX," let me point out that it amounts to the exact same thing (as the reader will see for himself upon reflection).

elements does, even so, have its validity. That validity is however restricted to the quantitative domain, which is to say that
the particulate model of SX enables us to calculate the measurable properties of SX to high degrees of accuracy. On the other
hand, contemporary physics comes in practice with an ontology
of its own: the ontology of Cartesian bifurcation, as we have
seen; and whereas the physics of SX may yield valid predictions, the associated ontology is nonetheless incurably flawed.
In this optic, SX is yet "the sum of its parts": of the very atoms
and molecules into which the physicist has decomposed SX.
But this decomposition, as we have seen, contradicts the fact
of irreducible wholeness, which stands at the very core of all
authentic ontology.

Our rejection of the scientistic ontology does not entail
a denial or repudiation of transcorporeal physics as such; it
does however cast that physics—namely, quantum theory—
in a brand new light. It first of all disqualifies the notion
that the macroworld of classical physics has emerged from
the microworld of quantum theory: what we have conveyed
apropos of IW negates that possibility categorically. It should
moreover be noted that this accords with the fact that the "big
bang" scenario of cosmogenesis has in truth been disqualified
empirically by the CMB data derived from the Planck satellite,
a matter which has not received nearly the acknowledgment
it demands.[19] The fact, in any case, is that the "macroworld" in
which we find ourselves did not—and indisputably *could not
have*—"emerged" out of a microworld consisting of quantum
particles. And let me add that the instant one has grasped
the Platonist ontology we have attempted to convey, this fact
becomes literally self-evident.

One sees namely that atomism of every description—from
that of Democritus to that of Stephen Hawking—has been
refuted once and for all on elementary ontological grounds: as

19. I have dealt with this question at some length in *Physics and Vertical
Causation*, op. cit., ch. 5.

we have noted, "particles" of whatever ilk, aggregated by whatever means, do not add up to "being," for the very simple reason that *being does not reduce to a sum of parts*! And if we are speaking of *quantum* particles in particular, there is—as we have likewise made clear—an additional point to bear in mind: the ontological fact, that is, that these putative particles *cease to be mere particles the instant they are actualized by contact with an IW.*

A categorical distinction needs therefore to be made between a *physical* atomicity—the kind, for example, that enables us to understand atomic spectra—and an *ontological* atomicity that would reduce subcorporeal entities to an aggregate of particles or atoms, howsoever conceived. The latter, I say, constitutes a claim which physics as such neither calls for, nor can justify, but can in fact be refuted on ontological grounds: our preceding considerations centering upon *irreducible wholeness* suffice to make this perfectly clear.

Meanwhile nothing obliges us to deny the truth of the contemporary atomistic "models"—so long as one realizes that the physicist's technical view of the matter does not give us the full picture. These atomistic components can do nothing *until they have become authentic parts of an irreducible wholeness*: only then *do they exist as bona fide parts of an actual entity.* What we affirm is that the requisite transition from *potency* to *act* can take place only by means of VC emanating from an IW derived from the corporeal plane—and thus, ultimately, from that "*pivot around which the primordial wheel revolves.*"

PART II

1

IRREDUCIBLE WHOLENESS
AND DEMBSKI'S THEOREM

We have become conditioned to think of wholeness in inherently set-theoretic terms, which is to reduce the whole to a sum of parts. There is a wholeness, however, which does *not* reduce to a sum of parts: an *irreducible wholeness* we shall say. Examples of IW are multitudinous and cover a vast spectrum of ontological domains. To begin with biology: whether our scientists have yet discovered the fact or not, every living organism—from the amoeba to the *anthropos*—is in truth an IW, which means not only that it does not reduce to a sum of parts, but implies that it cannot ultimately be understood on a "parts" basis as well. Very much the same can be said of a mathematical theorem or an authentic work of art, which likewise constitute IW's. It was Mozart who reportedly declared that "an entire symphony comes into my mind all at once," which of course needs then to be "unfolded" into an assemblage of notes so that the rest of us can apprehend it too. The point is that it is not the notes that make the symphony, but it is the symphony, rather, that determines the notes.

It proves however to be the rationale of our fundamental science—physics namely—to break entities conceptually into their smallest spatio-temporal fragments and thenceforth identify them with the resultant sum. Our very conception of "science"—of rationality almost—entails the reduction of wholes to an assembly of parts. One might say that the implicit denial of *irreducible* wholeness has virtually become for us a mark of enlightenment. It may therefore come as a surprise that mathematics—the most rigorous science of all—is in fact admissive of IW to say the very least, to the point that its formal exclusion from the discipline has required the collaboration of leading thinkers over a period of roughly three centuries. The project was initiated by René Descartes in the seventeenth when he "arithmetized" geometry through the invention of what to this day is termed a "Cartesian" coordinate system, and completed, if you will, in 1913 by Bertrand Russell and Alfred Whitehead with the publication of their august treatise entitled *Principia Mathematica*—read by only a stalwart few—that would reduce mathematics to a formalism in which IW has no place.

We need first of all to be apprised of the fact that not only in the days of Pythagoras and Plato, but in the premodern world at large, a triangle or a circle, for instance, was by no means viewed as a "point set," but was indeed conceived as an irreducible whole. To regard these geometric entities as mere aggregates would have been seen as a denial of their essence, their very "being." The same can be said of non-geometric entities such as integers or ratios thereof, which were likewise conceived in premodern times as IW's—even as they are in the context of musical scales, where one speaks, for instance, of the "octave."[1] By the time one arrives at what we term "real" numbers, on the other hand—the kind represented by nonterminating decimals—the picture has changed. To refer to these indiscriminately as "real" is to ignore the time-honored distinction between *rational* and *irrational* numbers: those which constitute integer ratios and those which do not. But why does that matter? It matters because therein, I surmise, resides the distinction between irreducible and reducible wholeness in the numerical domain.

Of course this calls for a good deal of explanation, which we shall attempt to suggest, at least, in the following section.

1. Their designation as "rational" numbers may thus prove to be more than a mere linguistic accident.

But granting that it is true: what difference does it make? The answer to this question, I believe, is simple: to amalgamate "rational" and "irrational" numbers into a single category of so-called "real" numbers is in effect to amalgamate irreducible and reducible wholeness, thereby depriving arithmetic of its innate *ontological* significance. This step alone bears witness to the fact that, in the post-medieval world, the traditional ontological wisdom of mankind has become a closed book: a "brave new intellectuality" has taken its place. Mathematics had by then become disassociated from its "irreducible" content to the point that *Principia Mathematica* could be seen as definitive of what in truth it is. The very notion of IW having faded into oblivion, the way was open for the reduction of mathematics to formal logic and set theory, thereby diminishing its content ontologically to the status of a nonentity. One might say that set theory is to mathematics what quantum theory is to physics: it is what remains when every last vestige of "being" has been exorcised.

Yet whether we recognize the fact or not, now as before mathematics deals essentially with *irreducible wholes*. Descartes notwithstanding, first of all, the ancient distinction between geometry and arithmetic still carries weight: *qua* IW, a triangle or a circle, for example, does not in truth reduce to the domain of numbers. Whatever computational benefits the amalgamation of these disparate mathematical realms may have achieved, we have surrendered the ontological insights their separation entails. It is one thing to "do" mathematics, and quite another to understand what, from a metaphysical point of vantage, mathematics is "about." My point is that when it comes to that other side of the enterprise, the recognition of geometry as an irreducible discipline proves to be a *sine qua non*. We have been far too quick to stigmatize the ancient savants as "primitive" and the like, when in fact there are grounds to wonder whether, meta-physically speaking, the shoe may not actually be on the other foot. For my part, I am persuaded that when it comes to the ontology of mathematics—to the recognition of IW, namely, as the essential—the ancient savants apparently discerned what we no longer comprehend.

Not only, thus, did they recognize the ontological distinction between an irreducible whole and a mere aggregate, but what strikes me as even more significant: they realized apparently *that it matters*! This is, after all, why the Pythagoreans were vis-ibly impacted by their discovery that the ratio of a side to the hypotenuse of an isosceles triangle is what we term an "irra-tional" number. Might there be then, conceivably, a connection between "irrational" in this mathematical sense and its broader connotation as a deficiency or lack of some kind, be it cognitive or ontological? Could the copresence of the *irrational* signify perhaps that the sensible world consists, in the final count, not only of irreducible wholes—of *being*—but of *"nonbeing" as well*?

Granting that the Christian ontology does not coincide with the Platonist but ultimately transcends it,[2] I would point out that the aforesaid surmise accords with Christian sources. Recall, for instance, the words of St. Augustine in the *Confes-sions*, where—addressing himself to God—he declares:

> I see these others beneath thee: an existence they have, because they are from thee; yet no existence, because they are not what thou art.

I am not suggesting, of course, that St. Augustine is speaking of irrational ratios: what he affirms unequivocally, however, is that the cosmos is made up not simply of "being," but entails per-force an element of "nonbeing" as well. On the yet higher level

---

2. On this question I refer to *The Vertical Ascent* (Philos-Sophia Initiative Foundation, 2021), ch. 12.

of Scripture, moreover, there is a well-known citation which makes the very same point: I am referring to "the ontology of the Burning Bush" enunciated in Exodus 3:14. Recall the scene: Moses asks God to declare to him His name, and is told in response "*Ego sum qui sum*"—which can be rendered "*I am that which is.*" It is not hard to see that this concurs essentially with the words of St. Augustine.

Getting back to the Pythagoreans: I presume the disciples of Pythagoras did in fact understand the rational/irrational dichotomy in a cognate key—which explains, among other things, why the words "*Let no one ignorant of geometry enter here*" were reputedly inscribed over the portal of Plato's Academy.[3]

There can be no doubt, on the other hand, that the modern world is decidedly—and one might say, blissfully—"ignorant of geometry." In obliterating the ancient distinction between geometry and arithmetic, and uniting *rational* and *irrational* numbers into a single category—labeled "real" no less—we have in effect formally eradicated the distinction between *irreducible* and *reducible* wholeness: between "being" and "nonbeing" in the mathematical realm. We have thereby closed the door to an ontological comprehension of mathematical science. Furthermore, having excluded "being" from the realm of mathematics, we have excluded it *ipso facto* from the physical sciences, and thus from the resultant *Weltanschauung*. Think of it: what has

3. For the seriously interested reader I would recommend two references which prove to be a goldmine of information regarding pre-Enlightenment mathematics, of which moreover we presently stand in dire need. On the side of geometry I refer to *The Philosophical and Mathematical Commentaries of Proclus on the First Book of Euclid's Elements* (Trans. Thomas Taylor; London, 1788), and on the side of arithmetic to the monumental classic by Albert Freiherr von Thimus, *Die Harmonikale Symbolik des Alterthums* (Köln, 1868). I might also mention the works of Hans Kayser based on that of von Thimus, which unfold these ideas as they apply to various domains, from music and architecture to the shape of leaves or of a violin.

been jettisoned is the very conception of IW: the very conception of *being* itself!

I find it ironic that the so-called Enlightenment has, in a way, replaced the *rational* by the *irrational*. And let me add, parenthetically, that this explains why "evolution" has become the dominant myth of our age: where *being* reduces namely to the sum of its parts, it originates evidently through an aggregation of particles—which is, after all, precisely what the tenet of "evolution" reduces to ontologically. A treatise chronicling the transition—the descent, actually—from Pythagoras and Plato to Darwin and the *Principia Mathematica*, conceived in light of the *rational/irrational* dichotomy, might prove to be enlightening. Suffice it to note that the decisive shift began apparently near the end of the Middle Ages with a radical denial of IW in the form of an enchantment with *nominalism*. In the wake of the Enlightenment, moreover, this negation—which initially was opposed to the intellectual and cultural mainstream—became the implicit credo of the modern age—its religion almost, one might say—at least in the Western world.[4]

The idea of irreducible wholeness is closely associated with the ontological notion of the "tripartite cosmos," which I take to be not only normative but "perennial" in the sense that, in a way, it has always been known.[5] According to this cosmography the integral cosmos divides "vertically" into three ontological domains: the *corporeal*, subject to both space and time; the so-called *intermediary*, subject to time alone; and ultimately

---

4. The reader interested in the cultural implications of this transition might wish to consult the chapter titled "'Progress' in Retrospect" in *Cosmos and Transcendence* (Philos-Sophia Initiative Foundation, 2021).

5. See especially *The Vertical Ascent*, op. cit., chs. 2 and 9.

the *aeviternal,* subject to neither.[6] One may conceive of this tripartite ontology in terms of a circle in which the center represents the aeviternal domain, the circumference the corporeal, and the interior the intermediary. I am persuaded that this representation is genuinely iconic, and can in truth be employed—somewhat like a mathematical formula—to draw conclusions: *ontological* conclusions to be precise.[7] And I regard this "icon" as an invaluable key to many questions, including the enigma of irreducible wholeness: a way of "seeing" what stands at issue.

The icon bears witness, first of all, to what may be termed the *ontological primacy of irreducible wholeness.* One "sees" that inasmuch as an IW does not reduce to the sum of its spatio-temporal components, it pre-exists on the aeviternal plane: one might say that it manifests an aeviternal prototype. This conclusion cuts of course against the very grain of contemporary thought, which locates *being*—or better said, what is left thereof—on the spatio-temporal plane. To be sure, to minds steeped in the *Zeitgeist* of our age, the very idea that there may be something "beyond" the spatio-temporal smacks of the unbelievable, the utterly fantastic. Add the notion that this "transcendent" and indeed "aeviternal" element constitutes the core reality supportive of spatio-temporal phenomena as such—and chances are not many will be left standing to hear you out.

Yet to the extent one is willing and able to assimilate this ancient—and as I suppose—perennial teaching, one begins to realize not only that it actually makes sense, but that it may well be in essence *the only ontology* that does. The problem is that we tend, almost irresistibly, to "spatio-temporalize" whatever

---

6. I would point out again that this tenet suffices in itself to disqualify Einsteinian relativity in both its special and general forms: the existence of the intermediary realm entails namely a globally defined simultaneity which rules out the possibility of an Einsteinian space-time. On this issue I refer to *Physics and Vertical Causation* (Angelico Press, 2019), ch. 5 and pp. 107-11.

7. Ibid., pp. 105-8.

presents itself to our view; in its present state, at least, our mind appears to be incapable of apprehending unmediated whole-ness. It is hardly surprising, thus, that philosophy in the authen-tic "pre-Enlightenment" sense demanded a stringent discipline: that there was perforce a "yogic" side to the enterprise. We read in the ancient books that the disciples of Pythagoras, for instance, were obliged to observe a vow of celibacy, and that Socrates characterized philosophy as "the practice of death": can you imagine a contemporary professor of philosophy so much as utter such words? In those pre-Enlightenment times, philosophy—so far from reducing to mere "theory" or "spec-ulation"—was in essence a matter of *sight*, of *seeing*—not "*as through a glass, darkly*"—but actually "*face to face*": without a spa-tio-temporal intermediary, that is.

For readers respectful of the Judeo-Christian tradition let me note that, given what Genesis has to say regarding the *Fall of Adam*, this should actually come as no surprise. The point is that this prehistoric *Fall* has reduced human nature to what St. Paul terms a "*psychikos anthropos*" who "*knoweth not the things of God.*"[8] We need to recall that man is traditionally conceived as a *cor-pus-anima-spiritus* ternary, corresponding to the Pauline *soma-psyche-pneuma*. What is missing in the *psychikos anthropos* is thus the highest component of his tripartite being: *pneuma* or *spiritus*, namely. But this is precisely the faculty that enables man to know "*the things of God*," which in this context we may take to be the "things" pertaining to the aeviternal realm. This then, I conjec-ture, may be the *Eden* from which Adam was banished, an exclu-sion which evidently has been passed on to his progeny.[9]

---

8. 1 Cor. 2:14
9. I have suggested elsewhere that, in the case of certain saints, the original plenitude of human nature may have been at least partially re-stored, giving them a certain access to the aeviternal plane. I have argued that this may explain such feats of clairvoyance as have been witnessed, for instance, in the case of Anna Katharina Emmerich, who appears to have perceived events pertaining to the distant past as well as to a future century. See *The Vertical Ascent*, op. cit., ch. 10.

~

Irreducible wholeness, then, is not an abstraction—not just somebody's theory—but the very *being*, rather, of existent things. In corporeal entities it is the *being* proclaimed by the spatio-temporal parts: the *one* within the *many*. *Being* proves to be inseparable from unity or oneness: *"Ens et unum convertuntur"* say the Scholastics. But how do we know that there *is* such an *ens*, that there *is* such an *unum*? The fact is we *do*; we know it as the *"whatness"* of the thing: as the *what* that it *is*.

It has thus become apparent that, strictly speaking, our physical science does not deal with *being* at all, that in fact it can have no inkling either of *ens* or *unum*. It cannot therefore comprehend even the corporeal, but is constrained to deal with mere *potentiae* actualizable in principle on the corporeal plane. But if, by way of physics, one is unable to conceive even the lowest tier pertaining to the tripartite cosmos, what to speak of the aeviternal plane, the source of IW!

A number of basic ontological recognitions are implicit in what has been said. It is to be noted, first of all, that the causality accessible to physics—which breaks naturally into a deterministic, a random, and a stochastic kind[10]—is based upon spatio-temporal parts, and acts upon spatio-temporal parts in turn. To the extent that it may be productive of wholeness, this part-based causality—which I designate by the adjective *horizontal*—gives rise evidently to a wholeness which itself reduces *ipso facto* to the sum of its spatio-temporal components. One arrives thus at what might be termed an ontological theorem:

> (*) *Horizontal causation cannot give rise to irreducible wholeness.*

10. "Stochastic" causality is a combination of the deterministic and the random kind, as exemplified for instance in Brownian motion, consisting of deterministic trajectories interrupted by random impacts resulting in an abrogation thereof.

Let me interrupt this train of thought to point out once again how utterly incongruous, from a Platonist point of vantage, the contemporary dogma of "evolution" proves in fact to be: it presupposes, after all, that organic wholes arise from a reordering of parts in antecedent organic wholes. It postulates, therefore, that wholes derive from parts and pieces by way of a causality itself derived from parts and pieces—which is either to deny the irreducible wholeness of living organisms, or to suppose that this wholeness can be produced by horizontal causation. It is to be noted thus that (*) suffices to disprove Darwinian evolution.

The question presents itself whether there exists a causation that *can* produce IW; and it is not hard to see that in fact there *must*—given that horizontal causality comes into play only on the corporeal plane, and therefore constitutes a *secondary* mode. What, then, is the primary? What else can it be than a causation arising—not from parts or from a wholeness reducible to parts—but from the primary wholeness that antecedes both spatial and temporal division! It is perforce this causality originating on the aeviternal plane—which I designate by the adjective *vertical*—that bestows the *being* and *oneness* upon a corporeal entity which render its wholeness irreducible. One arrives thus at a second conclusion:

(**) *It is vertical causation that engenders irreducible wholeness.*

It is to be noted that whereas VC originates on the aeviternal plane, it acts perforce upon the two lower strata, beginning with the intermediary, where it gives rise to what Aristotelians term *substantial forms*. What proves to be of capital importance is that these substantial forms are endowed with a capacity to exercise a VC of their own, which they can do inasmuch as, *qua* IW, they pre-exist on the aeviternal plane. One consequently arrives at a third recognition:

(***) *Vertical causation not only gives rise to substantial forms, but can originate from substantial forms as well.*

Let us not fail to observe the radical departure of this Platonist ontology from our current "scientific" outlook: the fact, first of all, that instead of the world being spatio-temporal, it is "time and space" that constitute a fragmentation of an inherently aeviternal world. And needless to say, that Platonist reversal cannot but impact our understanding of science to the point of negating our post-Enlightenment *Weltanschauung* in its entirety.

Finally, it behooves us to recognize the remarkable asymmetry between that ancient outlook and the contemporary. Fundamental questions, first of all, which appear to be virtually insoluble in contemporary terms, turn tractable in a trice when viewed in a Platonist optic: just think of the multitudinous phenomena associated with VC—*nonlocality* for instance—which to the physicist prove incomprehensible. Notwithstanding its vaunted immensity, moreover, there is yet something "puny" about our "brave new universe," in which there are *in fine finali* no frontiers worthy of human conquest. The very opposite can be said of the Platonist cosmos, a distant glimpse of which suffices to ennoble our life. What indeed are billions of years and light-years—supposing they exist—beside the tripartite cosmos, which dwarfs the entire spatio-temporal world, first by its "intermediary" realm, and ultimately by the "aeviternal," which transcends the normal compass of the human mind.[11]

∼

I wish now to point out that—in 1998 to be exact—IW has made its appearance in the form of a mathematical theorem of truly epochal significance. It has however done so incognito inasmuch as the theorem is formulated in the context of

---

11. At the risk of speaking what can only be *"foolishness to the Greeks,"* let me note that from a Christian point of vantage one sees that the spatio-temporal world as such is in a sense post-Edenic, and our confinement therein the result of what theology terms "original sin."

information theory, a discipline which conceives of "information" in inherently set-theoretic terms. Yet, implicit though it be, the idea of IW enters as the very crux of the theorem. This fact becomes manifest—not, to be sure, in terms of information theory—but on ontological grounds.

I am referring of course to the famous theorem discovered by the mathematician and information theorist William Dembski,[12] and associated from the start with the notion of "intelligent design." I shall argue that Dembski's theorem can in fact be seen as a special case of the ontological proposition (*), which, as you recall, simply states that *horizontal causation cannot give rise to irreducible wholeness.* What Dembski's theorem actually deals with, on the other hand, is not IW as such, but *complex specified information* or CSI, which can be no more than an information-theoretic instantiation of IW.

Yet the fact remains that the significance of Dembski's theorem can hardly be overstated: it invalidates, after all—in a single mathematical stroke—the mechanistic worldview that has dominated Western civilization since the Enlightenment. But in so doing, it raises a decisive question of its own: if horizontal causation cannot produce CSI, what is it, then, that can? Dembski and his colleagues seem to have opted for the notion of "intelligent design." Yet one sees, on the basis of (*), that this does not get to the heart of the matter. The fact is that Dembski's theorem demonstrates the existence of a causation which does not fit into our "flat" cosmology. This hitherto unsurmised causation proves to be none other than what I had termed *vertical* causality in the context of the quantum measurement problem.

The recognition implicit in Dembski's theorem turns out thus to be—not information-theoretic—but primarily *etiological.* It proves that wherever you encounter CSI—which is just about everywhere—you are confronted by an effect of VC.

12. *The Design Inference* (Cambridge University Press, 1998).

What ultimately stands at issue, however, is not CSI, but IW: *irreducible wholeness* namely, which is something incomparably more general. In the final count *it is IW that testifies to VC.* CSI enters the picture only by virtue of the fact that, in the context of information theory, it exemplifies IW. It needs to be understood that not only is the conception of IW ontological, but that it pertains in truth to the deepest ontology of all: i.e., the Platonist, which situates that primary wholeness on the aeviternal plane. But there is, strictly speaking, no common measure between CSI and IW.

From a metaphysical point of vantage then, Dembski's theorem affirms precisely that *it takes VC to produce IW*; the fact that horizontal causes cannot produce CSI constitutes a drastically special case.

∾

What is it, then, in Dembski's formulation, that can be identified as the information-theoretic counterpart of "irreducibility"? We must remember that what the ontological theorem excludes is not the production of a wholeness, but of an *irreducible* wholeness, precisely. How, then, can the notion of "irreducibility" be formulated in information-theoretic terms? It is here that a touch of genius comes perforce into play.

In keeping with the principles of information theory, Dembski configures his theorem in probabilistic terms. He envisages an "event space" Ω endowed with a so-called probability measure P, which to every measurable subset E of Ω assigns a real number P(E) between 0 and 1. The subset E is termed an "event," and P(E) is its probability. What is needed now is some additional stipulation which elevates E above the status of an "ordinary" event by rendering it—in Dembski's terminology—*detachable*. The very word itself suggests that a "detachable" target must be "more" than a mere point set: how otherwise could it be "detached." What is it, then, that can render a subset to be "more" than a subset: that is the question, which tends to be disheartening—until, that is, the idea strikes that what is actually "more" than just a subset is in fact an *intelligible* subset, to put it in Platonist terms. What could it be, then, that can render a subset "intelligible"? Well: a pattern, for example, which can be specified by a rule of some kind—a *specification*, Dembski calls it. And this works!

But not until a second condition is imposed: for it is evident that *any* nonempty target can in fact be "hit by chance." To rule out this possibility requires therefore another condition: what Dembski terms *complexity*. One needs to assume that the probability of our detachable event E is small enough to preclude "accidental hits." But whereas this stipulation is both necessary and sufficient to validate Dembski's theorem, it serves only to neutralize the probabilistic setting of information theory. Inasmuch as that probabilistic setting is itself extraneous from an ontological point of view, the function of complexity, if you will, is simply to return the issue to ontological ground. In this optic Dembski's theorem is seen to be inherently ontological, as we have claimed.

What horizontal causation cannot produce is an event that is both *complex* and *specified*. An example may suffice to convey the idea in nontechnical terms: if Ω consists of sequences of length 1000, made up of H's and T's—the outcome, say, of tossing a coin 1000 times—the stipulation that H's and T's alternate would constitute a specification: an exceedingly simple one, to be sure. What is actually "detachable," moreover, are not the H's and T's, which are elements of Ω, but the pattern, precisely. And let us not fail to discern the fact—as subtle as it is crucial—that the pattern as such is indeed "more than the sum of its parts," and thus constitutes an *irreducible whole*. Everything hinges upon this razor-sharp point!

It is to be noted that whereas the "parts"—the H's and T's—are visually perceptible, the "pattern," strictly speaking, is not. Beyond a certain threshold of complexity, at least, it is evidently possible to perceive the H's and T's in the given order without recognizing the pattern, which in Platonist parlance is termed an *idea*. There is consequently a fundamental distinction between the "sensible" and the "intelligible," the "seeing" of which calls for different faculties, corresponding to the distinction between *psyche* and *pneuma*.[13] The salient point is that, in light of the Platonist ontology, the "intelligible" is actually situated "above space and time," and thus *on the aeviternal plane itself*. Unbelievable as it may strike the contemporary mind, the efficacy of what Dembski terms a "specification" resides in the fact that its ultimate referent proves to be *aeviternal*.

∾

It may be worth noting again that I originally came upon "vertical" causation in the context of the quantum measurement problem, where it plays the decisive role. What renders the so-called "collapse of the wave function" mystifying is that horizontal modes of causality are simply not up to the task: it requires VC to effect this "collapse," which in truth it accomplishes, not over an interval of time, however short, but *instantaneously*.[14] This very instantaneity—i.e., that the action of VC is indecomposable—can itself be seen as a signature of

---

13. We have previously alluded to "the expulsion from Eden" as signifying the effective loss of *pneuma*. This needs however to be qualified. In his present state, man does still have the use of *pneuma*, but only as associated with *psyche*. And this is presumably the reason St. Thomas Aquinas classifies "intellect" (*pneuma*) as a faculty of the soul (*psyche*). What has been lost is not *pneuma* as such, but what might be termed the *pneumatic* vision: the faculty of sight on the aeviternal plane. In apprehending what we have termed the "intelligible," we do in fact apprehend something pertaining to the aeviternal plane, but we apprehend it "*as through a glass, darkly*" but not "*face to face*." Here *psyche* serves as the "glass" through which we see.

14. See *Physics and Vertical Causation*, op. cit., pp. 26-9.

irreducible wholeness. Following this initial recognition of VC in the context of quantum measurement, moreover, the action of VC could be detected in various domains of scientific interest, ranging from what physicists term *nonlocality* to biological phenomena which are likewise inexplicable in "horizontal" terms.

But whereas VC was initially identified by its "instantaneity," it has since become apparent that this is the result of a still more basic fact, which we henceforth take to be the defining characteristic of VC: i.e., that even as horizontal causation is the causality emanating from parts, so VC is the causality that *emanates from aeviternal wholes.*[15] Our theorems (*) and (**) are reflective of this fact.

A question obtrudes at this point: having characterized, in (**), the action of VC as productive of irreducible wholeness, one is hard pressed to conceive of "wave function collapse" as an effect thereof. Let us, then, consider this issue. A corporeal measuring instrument M acts upon a quantum system described by a wave function ψ. What the act of VC accomplishes on the corporeal plane is a change of state in M: a pointer, say, is moved to a certain position.[16] But this evidently does not affect the IW of the instrument, which remains unchanged. How, then, can it be said that the action of VC is productive of IW? To recognize this, one needs to look at what happens to the wave function ψ, which as one knows—or, in any case, should know—represents an ensemble of *potentiae.* And now we see the picture: the vertical act in question has been "productive of irreducible wholeness" in the sense of bringing a given *potentia* associated with ψ into an IW, an irreducible wholeness—that namely of the instrument M!

15. See *The Vertical Ascent*, op. cit., chs. 2 and 9.
16. It seems natural to suppose that the VC in question derives from the substantial form of the corporeal measuring instrument in accordance with (***), but this proves to be irrelevant.

∽

The fact is that, near the end of the twentieth century, VC has made its appearance as a hitherto unrecognized scientific reality—and this unquestionably signals the end of the present era and the imminent commencement of another. To what extent VC will permit itself to be "domesticated" through incorporation into a scientific framework of some kind, time will tell. One thing, at least, we know beyond any doubt: the hegemony of horizontal causation—and thus of physical science—has been broken irretrievably. It strikes me as a safe prognosis, moreover, that our outlook regarding the "ancient superstitions" of mankind may undergo a drastic revision. I find it not inconceivable, even, that somewhere down the road there may be a resurgence of interest in Pythagorean geometry!

2

SUBCORPOREAL PHYSICS
AND VERTICAL CAUSATION

WITH THE DISCOVERY of quantum mechanics in the early decades of the twentieth century, it seemed that physics had at last discovered its fundamental laws. As for the pre-quantum physics—thenceforth referred to as "classical"—this came now to be seen as a science of the macroscopic, dealing ultimately with ensembles of so-called quantum particles. The ground for such an interpretation had in fact been prepared by the kinetic theory of gases, which demonstrated how a macroscopic physics could be derived statistically from that of an "atomic" substrate. The belief was rife that the physics of the sense-perceived or "corporeal" world could likewise be derived from quantum theory on a statistical basis.

The one fact that might have caused doubt in that regard was the ironclad dictum of the Copenhagen school—headed by Niels Bohr and Werner Heisenberg—insisting that in the scenario of measurement, the instrument that measures a quantum system cannot itself be conceived in quantum-mechanical terms. That Copenhagenist postulate, moreover, has been from the start a source of astonishment and displeasure to the physics community at large; yet all attempts to eliminate or circumvent the offending dogma have ended in failure of one kind or another. It thus appears that this Copenhagenist *cut* is grounded

This chapter was co-authored with John Taylor, currently studying at University College London.

in an objectively real dichotomy of some kind, challenging the prevailing assumption that "at bottom" physical reality as such reduces to quantum particles.

This happens to be the very issue with which I dealt "from the ground up" in a monograph entitled *The Quantum Enigma*.[1] Approaching the matter from a metaphysical point of vantage, I concluded that there is—and must in truth be—an ontological discontinuity between the quantum system and the measuring instrument, and that in fact quantum particles—so far from constituting corporeal entities—reduce to mere Aristotelian *potentiae*, a view first proposed by Heisenberg himself. Distinguishing thus between what I term the *physical* and the *corporeal* domains, I argued that *the act of measurement entails an ontological transition from the former to the latter.* Inasmuch as such an ontological transition can only occur instantaneously, it follows moreover that the modes of causation known and knowable to physics—which I term *horizontal*[2]—cannot in principle account for such an effect. I was thereby led to recognize a hitherto unknown mode of causality—which I term *vertical*—distinguished by the fact that it acts *instantaneously*.[3]

It follows from these considerations not only that the ontological domain to which quantum mechanics applies is limited, but that it excludes the corporeal plane on which real qualities—e.g., colors—make their appearance. To be sure, this claim presupposes a realist view of visual perception, a matter with which I have dealt at considerable length.[4] The relevant point is that such a realist interpretation of the corporeal not only "exonerates" the Copenhagenist cut, but corroborates the

1. First published in 1995, it was reprinted by Sophia Perennis in 2005.

2. For the sake of clarity let us note that this "horizontal" causality operates in time by way of a transmission through space and may be deterministic, random, or stochastic.

3. By way of introduction to the subject of "vertical causation," we refer to chapter 2 of *The Vertical Ascent* (Philos-Sophia Initiative Foundation, 2021).

4. Ibid., chs. 5 and 10.

corresponding dichotomy on stringently ontological grounds. It is a matter of distinguishing categorically between what, strictly speaking, *is* and what as yet *is not*—the Heisenbergian *potentiae*—even if that which *is not* is *about to be*. And, on that basis, an "end of quantum reality" has been established once and for all.

Furthermore, it was noted in *The Quantum Enigma* that the ontological distinction between the corporeal and the physical realms entails a corresponding dichotomy, within the physical domain itself, between what I term the *subcorporeal* and the *transcorporeal* domains.[5] Meanwhile, physics itself has begun to weigh in on this issue. Within the last decade, namely, a group of physicists—headed by George Ellis—has begun to investigate the physics of the *subcorporeal*: of physical objects SX, that is, associated with a corporeal object X. And surprising as it may seem to the scientific community at large, that subcorporeal physics does not—*cannot* in fact—reduce to quantum theory! Inevitably, a thermodynamic system known as a "heat bath" enters into play. It is not difficult to realize, however, that a thermodynamics does not reduce to quantum theory: this follows from the Second Law, which affirms that the total *entropy*—thermodynamic system plus environment—cannot decrease. Time, therefore, has now a direction—a "future" and a "past"—which it does *not* have in a quantum-mechanical universe, where physical processes prove to be reversible. It thus turns out that *subcorporeal physics does not reduce to quantum mechanics.* Not only, then, is there a bound to the applicability of quantum theory, but that bound occurs *within the physical domain itself.*[6] In addition to the ontological cut between the corporeal and the physical domains, there exists a corresponding cut—within the subcorporeal realm itself—*between the*

---

5. A physical entity that is not subcorporeal is transcorporeal.

6. It is quite amazing, therefore, that to this day the physics establishment seems to regard the cosmos itself as something reducible to quantum theory!

quantum-mechanical and the thermodynamic domains. And needless to say: inasmuch as that discontinuity within the subcorporeal realm is expressive of an *ontological* dichotomy, *physics per se
cannot account for that cut.*

⁓

The presumption, in post-Enlightenment times, has been to
regard the *corporeal* as an effect of the *physical*: an epiphenom-
enon conceived presumably as a Cartesian *res cogitans* or "thing
of the mind." It now emerges, however, that this metaphysical
presumption is demonstrably false: so far from reducing to an
effect of the physical, we find that the corporeal impacts the
physical to the point of *affecting its very laws!* We propose to
prove, in fact, that *the aforesaid dichotomy in the physics of a subcorporeal object SX is effected by vertical causation emanating from
the corporeal object X.*

Let us note, first of all, that this recognition entails a complete reversal of our *Weltanschauung*: it literally turns our world
"upside down"—or better said, "right-side up," given that our
claim is justified. It is, after all, *being*—as opposed to *nonbeing* or
potentiae—that defines the iconic "above." The Enlightenment
has thus imposed an ontological inversion upon Western civilization: a rotation of 180 degrees if you will, interchanging the
"above" and the "below." One could in truth write a cultural history of the modern world based upon this geometric metaphor:
it happens, namely, that this geometric inversion is mirrored in
a cultural.[7]

Getting back to physics, the aforesaid ontological contention
can be formulated as a theorem:

The transition from quantum theory to thermodynamics on the

7. Cf. *Cosmos and Transcendence* (Philos-Sophia Initiative Foundation,
2021), ch. 7.

subcorporeal plane is effected by vertical causation emanating from a substantial form.

To be sure, the concept of *substantial form* is incurably onto-logical; the term itself pertains to the Aristotelian-Thomistic tradition which distinguishes between *substances* and *attributes*, both of which are given by *forms.*

The proof breaks organically into three parts to show respectively: (i) that a subcorporeal entity SX comprises a heat bath which is quantum-mechanical in the small and thermodynamic in the large; (ii) that this heat bath can only be effected by vertical causation; and (iii) that the vertical causation in question derives from the substantial form of X.

The first step of the proof has evidently been accomplished by the physicists who established the existence of a heat bath as an essential component of subcorporeal physics, and recognized that this imposes a limit to the application of quantum theory.[8] Apart from George Ellis, the premiere researcher in subcorporeal physics, reference should be made especially to Barbara Drossel, who—in a paper entitled "Ten reasons why a thermalized system cannot be described by a many-particle wave function"[9]—contributed substantially to the physics of the heat bath.

This brings us to the second step, which affirms that the implied transition from a quantum-theoretic to a thermodynamic system cannot be effected by horizontal causality. There are two evident

8. B. Drossel & G. Ellis, "Contextual Wavefunction collapse: an integrated theory of quantum measurement," *New Journal of Physics* 20, 113025 (2018).

9. *Studies in History and Philosophy of Science Part B*, vol. 58 (2017), pp. 12-21.

ways of recognizing this: first, inasmuch as horizontal causation operates *in time*, it cannot affect time itself—cannot, for instance, impose a direction upon a directionless time. It follows that the transition from a quantum system to a thermodynamic entity cannot be effected by horizontal causality, and must therefore be attributed to *vertical* causation. Alternatively, the conclusion follows by way of the "generalized Dembski theorem" from the known fact that *the heat bath does not reduce to the sum of its parts*,[10] which itself is implied by Barbara Drossel's findings.

The third step proves to be purely ontological, and can be dealt with on a Thomistic basis by distinguishing between a "creative" VC—Thomistically termed the "act-of-being"—and a VC derived from the substantial form of a corporeal entity.[11] It is clearly the latter kind that acts upon the associated physical object SX.

≈

Whereas the idea of "vertical causation" first imposed itself in the context of the measurement problem, it is to be understood that VC is no less universal than the *horizontal* causality with which physics is concerned. As I put it originally:

> The natural or "natured" world presupposes a creative or "form-bestowing" agency not simply in the sense of a first cause that brings the universe into existence, but as a transcendent principle of causality that is operative here and now.[12]

In terms of an integral ontology it can even be said quite rigorously that VC—literally "the causality of wholeness"—constitutes the primary causation, which is itself the cause of the

---

10. See "Irreducible Wholeness and Dembski's Theorem," ch. 1 above.
11. On the distinction between creative and substantial VC, see *Physics and Vertical Causation* (Angelico Press, 2019), p. 47.
12. *The Quantum Enigma*, op. cit., p. 109.

horizontal modes, and as such has power to override them.[13] And this is what happens—and *must* happen—in the process of quantum measurement: that is what was established "abstractly" in *The Quantum Enigma*, and what we have here corroborated, based upon recent discoveries in the physics of the subcorporeal.

We are referring to the recognition of the so-called heat bath as the definitive structure of the subcorporeal realm which enables us to understand the measuring scenario in a far more concrete and detailed manner. One sees that the ontological cut separating the *corporeal* from the *physical* has its analogue in the physics of the subcorporeal, where it manifests conceptually as a "cut" within the heat bath itself. This theoretic discontinuity is of course invisible to physics, the empirical fact being that "below" this demarcation quantum theory applies, whereas "above" a radically different physics comes into play.

It is worthy of note that this "invisibility" to physics of the cut itself testifies to the accuracy of physics as such, for it bears witness to the transcendent origin of that discontinuity. Whereas both "above" and "below" that invisible cut, namely, a rigorous physics—based of course upon modes of horizontal causation—is operative, "at the cut itself" no physics whatever applies. We find it striking how accurately this accords with the underlying etiology: for what VC effects is clearly the transition between the two domains. For a veritable "instant"—a *nunc stans* as the Scholastics would say— horizontal causation is superseded by a transcendent causality operative "here and now." We need, once again, to realize—as most assuredly the great philosophers of antiquity have—that the cosmos, manifesting itself to us sensibly as the corporeal world, comprises "higher" planes as well.[14]

What almost universally impedes the contemporary scientist

13. Cf. *The Vertical Ascent*, op. cit., ch. 9.
14. On this subject we refer especially to *The Vertical Ascent*, op. cit., where the tripartite nature of the integral cosmos—along with its implications for the sciences—are studied in depth.

from availing himself of these metaphysical recognitions is the evolutionist contention which predisposes him to believe *a priori* that things derive invariably "from below" through a more or less fortuitous aggregation of particulate components. Not only, however, is there no *bona fide* evidence whatsoever for this claim, but the very existence of *vertical* causation points precisely in the opposite direction. Take the heat bath, for instance: does it arise "from below" by some kind of evolutive process? That is a question we are at last in a position to answer: definitely not! We now know, in fact, on the basis of physics itself, that the heat bath is *not* brought into being by any process of temporal causality, but originates through an act of *vertical* causation which, being *instantaneous*, is as far removed from "evolutionist" as the human mind can conceive. The very "instantaneity" of vertical causation militates in fact against the "flat" cosmology of the evolutionist *Weltanschauung*, which simply has no room for anything "vertical"! It appears at this juncture that physics itself—applied to the subcorporeal plane—has opened the door to a rediscovery of the tripartite cosmology known to humanity prior to the Enlightenment, from the Platonist in the West to the Vedic in the East.[15]

---

15. The concept of the tripartite cosmos has been dealt with, first, in *Physics and Vertical Causation*, op. cit., and subsequently, in greater depth, in *The Vertical Ascent*, op. cit.

3

PONDERING THE BRAVE
NEW PARTICLE PHYSICS

FOLLOWING THE epochal discovery of quantum mechanics,[1] foundational physics has gradually morphed into the forbidding discipline of "particle physics." I say "forbidding" in a special sense: for not only is that post-QM physics incomprehensible to laymen, but it turns out ultimately to be incomprehensible to specialists as well.

We are grateful to Alexander Unzicker, that exceptional specialist capable of enlightening us. Besides first-hand familiarity with the workplace and a keen intelligence, he exhibits the rarest qualification of all: the courage, namely, to call things by their right name. My interest was aroused the moment when, upon opening his book,[2] I came upon the words: "There is no way to convince an expert that he or she has done nonsense for thirty years" (6). I realized instantly that this person has something to say.

Let me begin with a few facts concerning CERN, the European Center for "Recherche Nucléaire" located near Geneva, which actually plays a pivotal role in Unzicker's tale. Its fame rests largely on the fact that it houses the Large Hadron

---

1. One might date the event by the year 1925, in which Heisenberg and Schrödinger independently discovered the fundamental equation.

2. *The Higgs Fake: How Particle Physicists Fooled the Nobel Committee* (2013). References—given in the form of parenthetical citations—are to this book.

Collider, which is doubtless one of the wonders of the contemporary world. Here is what Unzicker tells us about it:

> The LHC has a circumference of 27 kilometers, it cost 9 billion dollars, its energy bill is like that of a major city, the protons cross the Swiss-French border, it is being paid for by 88 countries all over the world, the beam size is less than a needle... The underground beam, with the energy of a train, is 100 meters below ground, there are more than 5 million wires, it runs in an evacuated tube, and the temperature there is 'one of the coldest points in the universe.' (128)

Let this suffice as an introduction to CERN, which in truth is not simply a "center for nuclear research," but a formidable force in particle physics: a force to be reckoned with.

One begins already, perhaps, to sense what Unzicker is talking about when he tells us that "physics, after its groundbreaking findings at the beginning of the twentieth century, has undergone a paradigmatic change that has turned it into another science, or better, a high-tech sport, that has little to do with the laws of Nature" (5). But whereas, initially, this may sound excessive, I surmise the reader will concur wholeheartedly with this assessment before he is halfway through the book.

The central target of Unzicker's treatise is the "standard model" of particle physics, which to the well-attested "building blocks" consisting of protons, neutrons, and electrons, has added a motley collection of so-called particles said to have been detected by means of colliders such as the LHC. As the reader will discover soon enough, Unzicker is no fan of that prestigious theory: a "dumb, messy patchwork" he calls it, and charges not only that it comprises imaginary entities, but that its *modus operandi* violates basic scientific norms. In particular, he claims that "over the decades, high energy physicists have been hunting for ever rarer effects, just to declare as a new particle everything they did not understand." On top of that he charges that the culture associated with CERN "hides the basic problems physics ought to deal with" (6).

Unzicker maintains, first of all, that particle physics *à la* CERN fails to satisfy the condition of being "falsifiable"—in the sense of Karl Popper—traditionally regarded as a *sine qua non* of scientific methodology. "Be falsifiable!" he declares: "An hypothesis must predict something and leave the possibility of being *rejected* by observation if it is scientific fair-play and not ideology" (14). And he flatly maintains that "there isn't even a possibility to prove particle physicists wrong today" (36). After alluding to the complexities of the new *modus operandi* Unzicker goes on to observe:

> It is essential to understand (though high energy physicists never will) that at this point there is no objective interpretation, no unambiguous reality of an 'observed' experiment but an inevitable invasion of theoretical assumptions—in this case, of the grubby standard model… Please keep in mind that particle physics has long distanced itself from the experiment that acts as an impartial arbiter to judge theories. People today muddle along and don't even know anymore what is model and what is fact. (45)

Needless to say, this is no longer physics as understood and practiced right through the glory days of quantum mechanics. It appears that the benefits conferred by CERN have come at a cost greater even than so many billion dollars: the very culture, namely, upon which modern science had been based—the culture that brought forth an Isaac Newton and a Werner Heisenberg—has been effectively destroyed. The prevailing norms of high-tech physics—which one hesitates to call a "culture" at all—are not only different from those that presided over the rise of physics to what one can call its golden age, but prove to be indeed inimical. Above all, so far from satisfying the venerable requirement of being "falsifiable," the new physics exhibits a flagrant "unfalsifiability" as one of its salient characteristics. As Unzicker points out:

With respect to the standard model, particle physicists have developed this ludicrous attitude we are all used to: either we find a wonderful confirmation of our model or—even more exciting—a contradiction. In any case a tremendous success. This is the prototype of a nonfalsifiable model, according to Karl Popper—the opposite of science. (39)

As Popper himself noted long ago: *"If we are uncritical we shall always find what we want: we shall look for, and find, confirmations, and we shall look away from, and not see, whatever might be dangerous to our pet theories."* An apt description, it appears, of physics *à la* CERN.

~

There are other factors, of course, which likewise enter into play; I am thinking especially of what might be termed the philosophical formation. It appears to me that the old guard—from Planck to Heisenberg and Schrödinger say—derived their philosophical formation from two chief sources: from classical antiquity on the one hand, which in the early decades of the twentieth century was still respected,[3] as well as from contemporary critical schools such as logical positivism. Around the middle of the century, however, a new orientation began to manifest: pragmatism namely, as embodied by that American genius, Richard Feynman. Suffice it to say that Feynman had a major impact upon theoretical physics through his formulation of so-called "quantum electrodynamics" and the technique termed "renormalization" needed to make it work. I am not sure anyone actually understands either QED or renormalization; but the consensus, in any case, is that "it works." For the old school, on the other hand, that was not enough.

3. I might mention that my father's generation still read their Homer in Greek and Virgil in Latin in the European counterpart of what we term "high school." And there can be no doubt that this had its effect, that it made a difference.

In the earlier tradition of theoretical physics—as exemplified by the likes of Bohr, Heisenberg, Einstein, or Pauli if you
will—there was yet a philosophical culture of some kind, and in
the case of Werner Heisenberg, at least, a grounding in classical
philosophy, both Platonist and Aristotelian. The point is that,
by the second half of the twentieth century, that culture seems
to have disappeared. What came to replace philosophy properly
so called—from Plato and Aristotle to a rigorous logical positivism—appears to be a kind of pragmatism, as exemplified by
the practice of "renormalization." Here is what Unzicker has to
say on this question:

> The gradual takeover of that concept [renormalization], seen
> in a broad context, contributed to the transition from philoso-
> phy-based physics to the technical, math-recipe form that had
> become fashionable in the late 1920s. Paul Dirac, in an early
> intuitive statement, was particularly skeptical towards the 'com-
> plicated and ugly' theory called quantum electrodynamics. (75)

And Unzicker goes on to give a quotation from Dirac which,
in retrospect, strikes one as prophetic: "Some physicists may
be happy to have a set of working rules leading to results in
agreement with observations. They may think this is the goal
of physics. But it is not enough. One wants to understand how
Nature works" (75).

Well, it appears those days are over. One sees that the transition began in the late 1920's; as Unzicker explains: "Feynman's era
started when physics had to recover from the lack of orientation
caused by the quantum revolution" (76). By the time Feynman
enters upon the scene the ground had been prepared: by then
the emerging generation of physicists were ready to welcome the
"complicated and ugly" *modus operandi*, which ere long came to
dominate the field of particle physics. Here again is Unzicker:

> A technical, not to say superficial, way of doing physics gained
> ground, leading to a complete reorientation of physics in those

days. To be blunt, it was at that point that due to the lack
of true understanding, the collective displacement activity in
the form of feasible but unreflective calculations began to take
over. (75)

He then goes on to quote Helge Kragh, the biographer of Paul
Dirac, who makes the point with consummate precision:

> Improved in its details but not changed in essence—proved to
> be quite workable after all. The empirical disagreements became
> less serious, and by the end of the thirties most of the young
> theorists had learned to live with the theory. They adapted
> themselves to the new situation without caring much about the
> theory's lack of consistency and conceptual clarity... When the
> modern theory of renormalization was established after the war,
> the majority of physicists agreed that everything was fine and
> the long-awaited revolution unnecessary. (74)

To which Unzicker adds: "Kragh's comment precisely identifies
the beginning of the sickness that became today's intellectual
epidemic of particle physics. Back then they blew it" (76).

∿

Unzicker then proceeds to take aim at specific domains of contemporary particle physics, following which little, if anything,
remains standing. Having already cast serious doubts upon
QED, he turns to what is termed *chromodynamics*. Here is what
he has to say:

> A modern, though weird, variation of quantum electrodynam-
> ics is quantum *chromo*dynamics (QCD). Quantum chromody-
> namics is a paper tiger that by construction is unable to deliver
> results. The absurdity is usually wrapped in the term 'perturbative
> methods don't work here' and one needs 'nonperturbative meth-
> ods'... But nobody rejects the whole thing as nonsense. (77)

As non-specialists we have of course to take Unzicker's word on this issue. Yet questions arise: for if fundamental incongruencies can be swept under the carpet in a central domain such as QED, it is hardly surprising that "nonsense happens" in QCD.

Nor is Unzicker any more receptive to the lore of the stipulated *quarks*: "The idea of quarks does not explain anything" (89), he declares apodictically. To drive home the *ad hoc* nature of the quark hypothesis, he cites the discrepancies it was meant to explain. For example:

> Rather than predictions, the history of particle physics is full of unexpected problems. Just one of them was the size of the proton, which was inconsistent with the data of the Stanford Linear Collider (SLAC). As always in such cases, an *ad hoc* fix by means of auxiliary assumptions was invented in the form of 'gluons' (literally gluing the loose ends of the theory) and quark-antiquark pairs, so-called 'sea quarks'... A critic could easily assert that the sea quark and gluon components were simply *ad hoc* devices designed to reconcile the expected properties of quarks with experimental findings. (91)

When it comes to that mysterious property called "confinement," moreover, Unzicker is visibly unimpressed: "The so-called riddle of 'confinement,' why the three quarks in a proton cannot be separated, while indeed unexplainable in conventional terms, is just a self-imposed problem indicating that the quark model is baloney" (26).

Yet there is more to the confinement tale, for as Unzicker points out: "In the end, the quark model succeeded by the ironical trick of proving that no quark would ever be directly seen by a physicist. This liberated physicists from any need to demonstrate the existence of quarks in the traditional way" (102). It is likewise fortuitous that "there are no quantitative predictions of the quark model whatsoever" (104). Repeatedly Unzicker hammers the point: "This idea of quarks does not explain anything... It is precisely such fake understanding that, without being testable by a concrete observation, has eroded physics, the gradual spreading of the sickness being justified by the argument 'we don't have anything better'" (89).

Thus Popper wins again.

$\approx$

Summarizing the *modus operandi* which has engendered the exotic "particle zoo" of the standard model, Unzicker lays bare the underlying logic, if one may call it such. It happens, namely, that notwithstanding the *ad hoc* nature of the new particle physics—which proves to be its essential feature—there is yet a method, a methodology of sorts in play. When you boil it down—as Unzicker most assuredly does—you come upon an underlying strategy which in fact is rather simple. At the risk of speaking incomprehensibly to the uninitiated, here it is:

> That's how the system works: you assign a signal to one particle, A; then you do another experiment at higher energy where you remove all the A particles as background. It turns out that this does not describe the outcome, thus one baptizes the remaining signal as particle B. The next experiment, at higher energy again, removes As and Bs and their combinations as background. Call the remainder C. (114)

To boil it down even further, what cannot be understood in terms of known particles becomes *ipso facto* the discovery of a new one: a winning strategy, if ever there was one! In retrospect, moreover, one sees how radically counterproductive that erstwhile "principle of falsifiability" turns out to have been.

Adding to the excitement of the new physics are its horrendous magnitudes: the unimaginably small no less than the stupendously large; and both sides of the spectrum offer unprecedented opportunities for the discovery of new particles! To focus on the tiny:

Every little bump in the diagram of cross sections can be interpreted as a particle... However, physicists in that period started to classify particles regardless of their lifetimes—which were incredibly small sometimes, such as 10^{-25} seconds for the delta particle. This is methodologically absurd, because there is no way to get out of the collision point into any detector—the interactions of such particles remain totally theory inferred. (81)

It needs however to be noted that vital as these technical issues may be, there are social issues at play as well, which have impacted the "brave new physics" to at least a comparable degree. To start with, consider the fact that at the LHC "about 10,000 people are conducting *one* experiment," and that "the analysis as a whole has become impenetrable" (62). One wonders what would happen to the discipline of mathematics if the proof of a theorem, say, were broken into so many pieces, each assigned to a separate "team of experts": it does not take too much imagination to recognize that not only does such a "division of labor" engender problems, but more or less guarantees that these will be circumvented rather than solved.

The possibilities of "influencing" the outcome of an experiment are virtually endless, beginning with the fact that "some 99.99988 percent of the data are discarded after the collision because they are considered uninteresting in terms of the model assumptions" (42). As our guide goes on to point out: "If you have dozens of particles with hundreds of assumptions about their properties worked out in millions of impenetrable computer code, then there is plenty of stuff you can fiddle around with to make the outcome agree" (44). We need not belabor the point: the emerging picture proves to be amazingly clear—and so are the immediate consequences to be drawn therefrom.

The ambience of contemporary particle physics constitutes evidently an ideal environment for what is colloquially termed "politics" to manifest: "There is no field that is dominated in a similar way by hierarchical structures," Unzicker tells us. "It is obvious that opinions in high energy physics are homogenized by social and hierarchical pressure" (55). And in a section entitled "The Emperor's New Clothes," he goes on to say:

> One never sees divergent opinions published . . . all the papers are streamlined babble of several-thousand author collaborations... All the evidence suggests that the big detector collaborations are uniform sociological groups, incapable of conducting genuine scientific debate, with individuals incapable of expressing dissent. (56)

By way of ultimate summation, Unzicker turns to Shakespeare for the conclusive words: "*Though this be madness, yet there is method in't.*"

≈

Edging finally towards what he terms "the summit of absurdity"—the alleged discovery, that is, of the Higgs boson—Unzicker points out that this Nobel Prize-winning claim rests upon "the physical reality" of two other particles: the W and Z bosons namely. And to quote Shakespeare once more: "*Ay, there's the rub!*"—for it happens that the two antecedent "discoveries" are themselves by no means above suspicion. As Unzicker explains:

> The characteristic of the W boson was that it should decay into an electron and an (invisible) neutron after 10^{-25} seconds... Keep in mind that W bosons—like Z bosons—can never get into a detector, but are supposed to have a 'signature' that emits an (invisible) neutrino. Go figure. Thus all you need is the simplest particle in the world coming out of the collision, and something that is missing. This seeing by not-seeing is one of the most absurd developments of scientific methodology. (107)

After cluing us in on some other twists and turns related to the saga of the Higgs boson, Unzicker concludes:

> The practice of detection by non-detection had been fully established and automatized. After all, the W had to be there and had to be found sooner or later. Remember that no W would mean that the theory of electroweak interactions for which the Nobel was awarded in 1979 was wrong. (108)

There is some very suggestive material here somewhat reminiscent of a detective novel, which I would not wish to chop up. Suffice it to quote Unzicker's summary: "Ultimately, the top quark had to exist because the bottom quark needed a partner, as the W's and Z's had to exist because otherwise the standard model was wrong" (113).

It may be time for yet another quotation from the Bard: "*Something is rotten in the state of Denmark*" perhaps.

4

COSMIC VERSUS
MEASURABLE TIME

THERE IS A categorical distinction between space and time conceived, on the one hand, as cosmic bounds, and on the other,
as variables x and t to be determined by measurement. It is a
distinction, moreover, which the physicist is disposed to miss
inasmuch as he is committed to the latter conception: physics
is, after all, "the science of measurement." Yet the distinction
proves to be real: if in fact there were no "cosmic" time—which
as such is *unmeasurable*—there could be no time coordinates t
as well.

What, then, *is* that unmeasurable time? Strangely enough, it
is something which itself "measures" in a very ancient and ontologically decisive sense: what it measures or "metes out," namely,
are *events*. It measures an event—not of course by assigning a
number to its duration—but by defining its beginning and its
end. And in so doing it gives rise to the event, causes it in a
sense to exist. That beginning and end—so far from being mere
attributes—are rather constitutive of that event, even as the
spherical boundary of a billiard ball is constitutive of that entity.
In brief: *cosmic time gives rise to events through an act of "cosmic
measurement"*—which is the reason we refer to it as a *bound*.

It is to be noted that this conception of time as a "bound"
differs from the more customary notion of time as a "container
of events," the analogue to the "empty container" conception
of space. This is not to say that time has not also an "empty

container" aspect, or that space may not have likewise an active aspect of "bound." The point is simply that in speaking of *cosmic* as opposed to *measurable* time, we are speaking of time as an inherently active principle of cosmogenesis.

~

It behooves us now to recall the concept of the *intermediary plane*,[1] which in a way we all enter, for example, in the experience of dreams. It is hardly necessary to point out that objects perceived in a dream have no location in space: a dream castle has no spatial coordinates—which is of course why, upon waking, we regard it as "unreal." But even though dream objects are not spatial—have no location in space—*they prove nonetheless to be temporal*: that is the point. We have all presumably experienced the ring of an alarm clock interrupting a dream sequence at a moment which could indeed be identified within the sequence itself. The fact is that *the dream state is subject to the bound of time* even though its objects do not exist in the spatial or "corporeal" world. They pertain thus to an ontological stratum our sciences have left out of account, which we term the intermediary plane.

The contemporary reader is of course prone to "psychologize" that ontological realm: relegate it instantly to the limbo of *res cogitantes* or "things of the mind"—which is however to miss the point utterly. So far from reducing to something mental or imaginary, the intermediary plane is as real as the corporeal world—and in a way more real inasmuch as it precedes the corporeal ontologically. It stands to reason, namely, that the imposition of a spatial bound does not bestow reality, but in fact restricts, and therefore in effect abrogates a pre-existent plenitude. There is in truth nothing in the corporeal realm which does not pre-exist ontologically in the intermediary. The latter is consequently not to be conceived as something "additional" to

1. Cf. *The Vertical Ascent* (Philos-Sophia Initiative Foundation, 2021), pp. 13-5.

the corporeal: to restrict is not in truth to create something new.

It is crucial to regain clarity on these fundamental issues, which have been hopelessly misconstrued since the onset of what has been euphemistically termed the Enlightenment. It needs to be understood that whereas, in our culture, all reference to *higher* ontological planes appears "mystical" in the pejorative sense, this is due to the fact that our post-Cartesian *Weltanschauung* leaves no room for ontological height. An entire dimension—the *vertical*—has been, as it were, verboten. The objectively real world, according to René Descartes, is made up exclusively of *res extensae*, objects extended in space; everything else is supposedly a *res cogitans.* This is the cosmology that gained dominance during the seventeenth century, presided over the rise of physics, and to this day defines our ontological status quo: the problem is that it proves to be misconceived.

It should above all be realized that "extension in space" is by no means indispensable as a condition of objective reality: it turns out that *time is more basic than space*, and that in fact there exists an ontological stratum subject to time alone, from which the corporeal or spatio-temporal world is derived through the imposition of the spatial bound.

The reason we characterize the aforesaid domain as "intermediary" is that, if time is indeed a *bound*, there must be something it bounds or acts upon: and this entails the existence of a primary cosmic domain not subject to time. This ontological fact is recognized in the great metaphysical schools of antiquity, in light of which that primary domain has been characterized as the *aeviternal*.[2] I would add that the resultant conception

---

2. What distinguishes "aeviternity" from "eternity" *per se* is that the former can be conceived in a cosmic—as opposed to a purely metaphysical or theological—sense. As St. Thomas Aquinas has put it: what defines aeviternity is that "time can be adjoined to it."

of a cosmic trichotomy is apparently indigenous to the ancient metaphysical schools: we encounter that conception from Greece to India, where to this day one speaks of the *tribhuvāna* or "triple world." It is, in particular, basic to Platonism, and accords moreover with the tripartite conception of man as *corpus-anima-spiritus*.[3] I have found it enlightening to represent the tripartite cosmos by a circle in which the center represents the aeviternal realm, the interior the intermediary, and the circumference the corporeal; and whereas it is unclear whether this representation was employed in the ancient schools, it strikes me as a veritable icon of the cosmic trichotomy.

This ontological tripartition is however invisible to the physicist for the simple reason that he looks upon the cosmos through corporeal instruments. What he observes is consequently subject to the bounds of space and time, which signifies that the corporeal realm is all he can "see." Let us not fail to note, on the other hand, that by the same token he cannot conclude that the tripartite cosmos does *not* exist: as I have noted repeatedly, such a denial pertains not to science properly so called, but to *scientism*.

We part company, thus, with Bertrand Russell when he declares that "what science cannot tell us, mankind cannot know"—a dictum which proves moreover to be patently spurious inasmuch as science itself can evidently tell us no such thing. I maintain, furthermore, that a *bona fide* science of the intermediary realm is in truth possible, even though such a science cannot be based upon observation via corporeal instruments, the point being that *the scientist himself can in principle serve as the instrument of observation*. I shall in fact argue that the ancient sciences fall typically into this category.

It is of course to be noted that a serious inquiry into the nature of ancient sciences is hardly to be expected from our contemporary savants, who seem in fact to relegate these sciences *a priori*

---

3. These are matters I have dealt with at length in other publications, notably *The Vertical Ascent*, op. cit.

to the category of "ancient superstitions." By way of response, let me share with you an anecdote which bears very directly upon this issue. It was my first visit to India—which now goes back more than half a century—and I had just arrived in New Delhi, when I received word that a personage whom I was very eager to meet was scheduled to arrive the following morning at 11 AM. Highly pleased, I took a stroll through the city, and on my way back to the hotel was accosted by a fakir, who—out of the blue—informed me that something "very good" would happen to me "tomorrow at 11 AM." Since no one in New Delhi besides myself could know of this, I became interested. Repairing to a nearby garden with this individual, I wanted to put him to the test. He began by handing me a piece of paper, which he requested me to examine: I found that it was blank. He then asked me to fold the paper and hold it in my hand. Next he asked that I think of a number between 1 and 100. I did, choosing the number 36, as I recall. Then he asked me to examine the paper: and there, clearly visible, was the number 36.

What are we to make of this? I realized soon enough that the number 36 must have been inscribed on that slip of paper in so-called "invisible ink," which turns visible when it is warmed—as presumably happened when I held it in my hand. Yet that number, it appears, must have been conveyed directly from the fakir's mind to mine, *a transfer which I presume can only be situated in the intermediary plane*: that much is clear. It appears that this fakir was able to accomplish two things our Western savants believe not to be possible: first, to pick information directly out of someone's mind (the "11 AM"); and second, to reverse the process by putting something directly into another person's mind (the number 36).

What, then, *is* a "fakir"? Suffice it to say that he represents the lowest grade of a *yogi*: someone, that is, who has acquired some degree of proficiency in the ancient practice of *yoga*. The fact is, however, that this can only be done as a disciple of a qualified *guru*: there is in truth no such thing as a "do-it-yourself" *yoga*!

As for the fakir, to exhibit yogic feats of whatever kind for pecu-
niary gain renders him contemptible in the eyes of authentic
yogis, who most assuredly are in quest of greater things. Yet,
even so, that bit of "supernatural" know-how did not come
cheap: that fakir too was obliged to pass through a novitiate of
arduous practice under the tutelage of a qualified master. What
is more—and hard for us to grasp—is that "instruction" alone
does not suffice: what is also called for as a *sine qua non* for the
practice of authentic *yoga* is an initiation or *dikshā*, which itself
transpires on the intermediary plane.[4]

∾

I have related this anecdote to point out that there exist sciences
that enable its practitioners to transcend the corporeal realm,
which differ fundamentally from our own in that they not only
presuppose powers on the part of the practitioner which need
to be acquired by rigorous disciplines practiced over extended
periods of time, *but hinge upon an "initiatory" transmission*, an
authentic *dikshā*. An element unknown in the modern world
comes thus perforce into play: *discipleship*, that is. The sciences
to which we refer are therefore *traditional* not only in the sense
of being "handed down" from master to disciple, but also in that
they are *initiatory*. Only thus, namely, does a controlled pen-
etration into the supra-corporeal planes become possible: the
fact is that *to enter upon the "supernatural" one requires ultimately
"supernatural" means.*

To be sure, the origin of such "initiatory chains" is for us
mysterious in the extreme—which should however come as no
surprise: how could it be otherwise, given what has already been
set forth! The Christian reader, on the other hand, recalling the
Adamic Fall, needs hardly to be surprised. The fact, in any case,

---

4. The Catholic reader may discern a certain analogy—distant though
it may be—with the priesthood, which likewise hinges upon a chain of
transmission through the intermediary plane.

is that *empirical science can transcend the corporeal plane only to the extent that the scientist himself has done so*, which entails that a "religious" element—in the broad sense of *re-ligare* or "binding back"—enters necessarily into the picture.

As to our fakir who can pick information out of someone's mind as well as implant ideas therein: if the lowest grade of a yogi can perform such feats, what then is to be said of those who have devoted themselves fully to yogic practices over the better part of a lifetime? I have personally witnessed yogis seated motionless in deep meditation for most of their days and nights, and have sensed upon their return to normal consciousness that they came as it were from afar. I doubt not that their objective is to transcend the intermediary domain, which in keeping with Vedic tradition they regard as no more than a corridor to be crossed in quest of their proper destination. It appears moreover that even as there are means of horizontal locomotion, there are means of vertical "journeying," unknown though these be to contemporary civilization. It needs to be realized that there is in truth no common measure between the two: for whereas horizontal locomotion transpires within a given ontological stratum, the vertical brings you face to face with "things undreamed of in your philosophy."[5]

This brings us to the question whether there exist *bona fide* empirical sciences transcending the corporeal plane: and let me say that, for my part, I believe there are. I am fully persuaded, for example, that alchemy is in truth a case in point, and that its two basic operations—the *solve* and *coagula*—constitute indeed transformations from the corporeal to the intermediary and the reverse, respectively. Having caught glimpses—though for the most part vicariously—of higher ontological strata, I rank alchemy among the things "not dreamed of" in our present-day philosophy.[6]

---

5. The full quote from *Hamlet* being: "There are more things in heaven and earth, Horatio, than are dreamt of in your philosophy."

6. On the subject of *astrology*—another science which hinges upon higher ontological planes—I refer to *The Vertical Ascent*, op. cit., chapter 7.

It is time to break the confines of the Cartesian *Weltanschauung*: time to rediscover the integral cosmos, the world which cannot be measured in light-years. Our current Occidental provincialism was in a way excusable during the glory days of physical science, when each new triumph was followed by another more astounding yet; but now that fundamental physics has entered a state of manifest chaos,[7] the picture has drastically changed. Today the notion that the cosmos in its totality may comprise additional ontological planes—so far from being unthinkable—has in fact become eminently plausible: when the physics of so-called fundamental particles has turned into an ugly mess and no one admittedly understands[8] the one theory that actually works—it is hardly the time to reject alternative conceptions out of hand.

∿

The existence of the intermediary plane—or more precisely, the tripartite nature of the integral cosmos—has profound implications for physics, beginning with the recognition that inasmuch as causality originates on the aeviternal plane, the primary causation *must* in fact be *vertical*. What I showed in the context of the quantum measurement problem is that a causality operating *in time* cannot account for the collapse of a wave function; and what emerges now from the tripartite nature of the integral cosmos is that *the primary causality is perforce vertical.*[9]

It is evident that physics, as "the science of measurement," is restricted to the lowest stratum of the cosmic trichotomy: the corporeal plane. It deals thus with entities subject to both space

---

7. See ch. 3 above.

8. I am referring to Richard Feynman's justly famous assertion that "No one understands quantum theory."

9. Vertical causality—which operates not in time but "instantaneously"—is differentiated from what I term *horizontal* causality, which operates in time by way of a physical process.

and time, and "has eyes" only for *quantities*: what it "sees"—or better said, *can* see—pertains to the quantitative side or aspect of corporeal reality. The very fact, however, that this quantitative content arises from bounds imposed upon a pre-existent reality implies that *corporeal reality as such does not reduce to the categories of physics*. It follows that *physics is unable to know the corporeal world*, another recognition at which I had originally arrived in the context of the quantum measurement problem.

What is moreover invisible to physics are not only attributes—in particular *qualities*, such as color—but *substances* as well: it turns out, namely, that the very notion of *substance* has no place in "the science of measurement." What this ultimately entails is that, "*on its own level*" so to speak, *physics reduces necessarily to a quantum mechanics*; as Sir Arthur Eddington observed at the outset of the quantum era: "*The concept of substance has disappeared from fundamental physics.*" And I would note that the overwhelming emphasis upon what is actually *measurable*—an emphasis characteristic of early twentieth century philosophical movements such as logical positivism—has rendered that disappearance virtually inevitable. As in the case of qualities, the idea of "substance" had to go, not because *there is no* substance, but simply because it *is not measurable*.

In a word, "unmeasurables" of every description were officially relegated to the limbo of *res cogitantes* in a relentless campaign to render the cosmos fully "measurable." In addition to sensible qualities and substances, however, there are yet other unmeasurables which likewise "carried a price on their head." The most basic among these is evidently the distinction between the bounds of *space* and of *time*, which clearly does not pertain to the quantitative realm. And it was Albert Einstein who apparently accepted the task of its elimination as his personal mission: it is this that characterizes Einsteinian physics in its totality and separates it sharply from the non-Einsteinian.

The first thing to note, in this regard, is that the tripartite cosmology not only distinguishes categorically between space and time, but in fact *assigns ontological priority to the bound of time*. How, then, does this ontological priority impact physics: what physical consequence does it entail? It impacts physics, I say, by the fact that it entails *a globally defined simultaneity*, a notion incompatible with the Einsteinian reduction of time to space. This means that Einstein could achieve his objective only by formulating a mathematical physics in which *there is no global simultaneity*. And therein resides the endemic error of relativistic physics *per se*.

The recognition, on the other hand, that there does exist a universal cosmic "now" is consonant with classical (Newtonian) mechanics. To comprehend ontologically what stands at issue, let us reflect somewhat upon the act of imposing *spatial separation* upon a domain in which *there is time but not space*. It is clear, first of all, that this act must take place instantaneously: that it cannot, in other words, be conceived as taking place over an interval of time. But inasmuch as motion presupposes an interval of time, this in turn implies that what is thus imposed cannot itself depend upon motion. Now, even as the bound of time gives rise to *duration*, the bound of space engenders *distance*. It follows that *distance does not depend upon motion*, contrary to what Einsteinian physics affirms.

One sees, finally, that Einstein's error was not scientific, but philosophical: it was his unbending logical positivism that led him astray. Given the—*erroneous*—assumption that the unmeasurable does not exist, his work appears however to be impeccable.

∾

At this juncture I would like to interpose a brief digression referring back to earlier work.[10] The fact is that whereas the equations of mechanics transform according to the Galilean rule, the equations of electromagnetism do not: what does this

10. *Physics and Vertical Causation* (Angelico Press, 2019), ch. 5.

imply, what does it signify? In contrast to an Einsteinian Principle of Relativity, what stands at issue is actually a Principle of Immobility which singles out a special class of inertial reference frames as immobile or at rest: those, specifically, that are stationary with respect to the Earth. And shocking as it may strike the physics establishment at large, this fact provides a rigorous basis for a revalidation—yes, of *geocentrism* no less! That long-abandoned and much-maligned notion may yet replace Einsteinian relativity in the end.

Let me note in passing that the authentic "Einstein story" has yet to be told—and the more carefully one separates the facts from the myth, the more amazing that story becomes. Whatever facet of "relativistic physics," namely, one chooses to examine from an empirical point of view begins, upon close enough scrutiny, to disintegrate. The single exception, to be sure, is the fateful formula $E = mc^2$, which proves to hold its ground: the catch is that this formula has nothing whatsoever to do with Einsteinian relativity![11] I am personally persuaded that what might not inappropriately be termed the "Einstein mania" has from the start been driven by ideology: what ultimately stands at issue is the complete quantification of cosmic reality, which constitutes after all "the holy grail" of the scientistic enterprise.

In conclusion let me say that I nonetheless hold Albert Einstein in high esteem. The fact alone that in the year 1905, in which he inaugurated his special theory of relativity, he published two other papers—one on Brownian motion and one on the photoelectric effect, either of which is richly deserving of a Nobel prize—this in itself ranks him as a consummate master of his discipline. His shortcomings, therefore, reduce to the limitations of physics *per se*.

11. It can be derived from classical physics and is in fact on record in the nineteenth-century journal literature.

Getting back to the tripartite cosmology and the emergence of vertical causation, let me recall that three years following the publication of *The Quantum Enigma*—in which I introduced the latter notion—a mathematician by the name of William Dembski published an epochal theorem proving that wherever there is "design" manifesting as so-called "complex specified information" or CSI, it is an effect of VC. It follows, thus, that VC is operative not only in acts of quantum measurement, where it was first identified, *but in creative activity of every description*: in every process, that is, which results in the production of CSI. And this includes the better part of normal human activity, which in fact is human precisely by virtue of not being robotic.

The production of CSI is not however limited to conscious human endeavor, but is operative throughout the biosphere. What defines a living organism is a wholeness not reducible to the sum of its parts. It happens, moreover, that such an irreducible wholeness *is in principle productive of VC.*[12] And this recognition brings traditional metaphysical doctrine once again into play: let me explain.

That IW definitive of the biosphere proves namely to be none other than what is traditionally termed the *soul* or *anima* of the living organism; and *what renders that soul irreducible to the sum of its parts* is the fact that *it pertains to the intermediary domain*, and consequently *has no* (spatial) parts! The crucial point, now, is that whereas this *anima* is itself an effect of VC[13]—primary VC one can say, what St. Thomas Aquinas terms *the act-of-being*—it is yet capable of exerting a vertical causality of its own: it happens namely that an IW has

12. What stands at issue is an ontological alternative to Dembski's information-theoretic approach which vastly generalizes his theory. See ch. 1 above, "Irreducible Wholeness and Dembski's Theorem."

13. This follows from the generalized Dembski theorem, as per the aforesaid chapter.

in principle that capacity.[14] And *this VC emanating from the anima constitutes the vertical causation definitive of the biosphere,* what in times past was sometimes referred to as *élan vital.* The point is that a living organism does not reduce to a mechanism, and *cannot therefore be understood in terms of horizontal causation.*[15] Contrary to what Western savants have believed since the days of Galileo and Descartes, *biology does not reduce to physics.*

It appears that physicists as well as biologists engaged in fundamental research are beginning to realize this fact, which is to say that we are entering a new era in which a partial rediscovery, at least, of the ancient cosmological wisdom seems destined to take place. Notwithstanding the attendant uncertainty, the period known to historians as the *Enlightenment*—what René Guénon designates more soberly as "the reign of quantity"—is coming to an end.

14. I refer once more to ch. 1.
15. Cf. footnote 9 above.

INDEX OF NAMES

ABOUT THE AUTHOR

WOLFGANG SMITH was born in Vienna in 1930. At age eighteen he graduated from Cornell University with majors in physics, mathematics, and philosophy. At age twenty he took his master's degree in physics at Purdue University, subsequently contributing a theoretical solution to the re-entry problem for space flight while working as an aerodynamicist at Bell Aircraft Corporation. After taking his doctorate in mathematics at Columbia University he pursued a career as professor of mathematics at M.I.T., U.C.L.A., and Oregon State University until his retirement in 1992.

Notwithstanding his professional engagement with physics and mathematics, Wolfgang Smith is at heart a philosopher in the traditional sense. Early in life he became deeply attracted to the Platonist and Neoplatonist schools, and ultimately undertook extensive sojourns in India and the Himalayan regions to contact such vestiges of ancient tradition as could still be found. One of the basic lessons he learned by way of these encounters is that there actually exist higher sciences in which man himself plays the part not merely of the observer, but of the "scientific instrument": i.e., becomes himself, as it were, the "microscope" or "telescope" by which he is enabled to access normally invisible reaches of the integral cosmos. By the same token, Smith came to recognize the stringent limitations to which our contemporary sciences are subject by virtue of their "extrinsic" *modus operandi*: the folly of presuming to fathom the depths of the universe, having barely scratched the surface in the discovery of man himself.

Following his retirement from academic life, Smith devoted himself to the publication of books missioned to correct the fallacies of contemporary scientistic belief by way of insights derived from the perennial wisdom of mankind. These works focus primarily on foundational problems in quantum theory and the no less challenging quandaries related to the problem of visual perception. The key to the overall puzzle, according to Smith, is to be found in the long lost cosmology of antiquity, which conceives of the integral cosmos as tripartite—even as man himself is traditionally viewed as a composite of *corpus*, *anima*, and *spiritus*.

The Philos-Sophia Initiative Foundation has produced a feature documentary on the life and work of Dr. Smith, *The End of Quantum Reality*, which is available on disc and digital platforms worldwide. Visit theendofquantumreality.com for more information.

Lightning Source UK Ltd.
Milton Keynes UK
UKHW021958250123
415976UK00005B/113

9 798985 147032